I0034225

QUANTO-GEOMETRY

OVERTURE OF COSMIC CONSCIOUSNESS OR UNIVERSAL KNOWLEDGE FOR ALL

Volume I – Revision I

Joseph J. Jean-Claude

Quantometrix, Inc.

Worcester - USA

Publisher's Cataloging-In-Publication Data

Names: Jean-Claude, Joseph J.
Title: Quanto-geometry : overture of cosmic consciousness or universal
    knowledge for all. Volume I / Joseph J. Jean-Claude.
Description: Revision I. | Worcester, MA : Quantometrix, Inc., [2017] |
    Originally published in 2015. | Includes bibliographical references and
    index.
Identifiers: ISBN 978-0-9647466-5-7
Subjects: LCSH: Quantum theory. | General relativity (Physics) | Con-
    sciousness. | Cosmology. | Matter.
Classification: LCC QC174.12 .J43 2017 | DDC 530.12--dc23

© 2015 - 2018 QUANTOMETRIX, INC.
97 Webster St.
Worcester, MA 01603

All rights reserved. Printed and bound in the United States of America. No parts of this book may be reproduced or utilized in any form or by any means, electronic or mechanical, including photocopying, recording, or by any information storage and retrieval system, without permission in writing from the publisher.

International Standard Book Number: 978-0-9647466-5-7

Acknowledgements

To my son, Joel Jean-Claude,
who was kind enough to break his busy schedule for a pre-press reading of this text and to provide comments and a Preface which no doubt contributed to its betterment and completeness.

To my brother, Frantzy N. Jean-Claude,
for the endless late-night discussions over matter, philosophy and life, a continuing chapter of our lives and an inspiration.

PREFACE

This text is primarily aimed at researchers, scientists, physicists, and mathematicians who are intimately familiar with many of the long standing theories in modern physics, i.e. the Theory of Relativity, Big Bang Theory, etc. However, anyone who has covered courses in at least Calculus 2, Physics 102 and Chemistry 101 will intellectually benefit as well from the discussions carried out in this proposal.

Background in the referred level of Calculus and Physics courses is important since the analysis stems from these very root notions in the physical sciences. While knowledge of Quantum Mechanics (QM) and Quantum Field Theory (QFT) is not a requirement (the author does a fairly good job at succinctly explaining the basics when necessary), it will help take stock of the contrasts between Quanto-Geometric constructs and their parallel QM and QFT constructs whenever applicable.

The first two Chapters of this text represent a necessary introduction to the main theme, the derivation of fundamental physical constants from first principles. Primarily aimed at those with an undergraduate level understanding of physics, the discussion in those Chapters lays the foundation for and spells out those principles in a clear and granular fashion. The graduate audience may skip the first two chapters and directly start with Chapter 3, if they so desire.

Chapter 3 is the beginning of what some of the more novice readers might consider "the good stuff", where the author presents a new perspective on things learned in math in earlier cycle of education (formally known as Number Theory), and other curious things that one may have noticed along the way, but could never explain why. There is an intertwining of sorts between these math abstractions and the Quanto-Geometric principles.

Chapter 4 represents an alternative introduction to the main subject, especially written for graduates. The undergraduate audience may skip Chapter 4 and proceed to Chapter 5 from Chapter 3, if they so desire.

Chapter 5 and 6 are a showcase of theoretical discoveries and where the bulk of the contributions germane to this proposal are delivered. Speed of light constant: derived! Planck constant: derived! Gravitational constant: derived! Rydberg atomic orbital constant: derived! And so on and so forth. It is no less interesting to witness the unraveling of the notion of time, under assault by the author throughout the entire content, and its ultimate consequences as spelled in Chapter 6.

Chapter 7 proposes a new vision of statistical physics, fills in the gaps in the Quantum Theory which caused generations of theoreticians to tax it as analytically incomplete, while demonstrating that Statistical Theory always contained the promising seeds of the Grand Unified Theory.

All in all, this book should be informative, enlightening, engaging, and most of all, exciting: a complete breakthrough of crucial physics conundrums that have challenged generations of physicists since early 20th century. Let's begin.

TABLE OF CONTENTS

Chapter 1

To discover the higher Truth requires
the courage to put everything on the table,
at least for once, to whatever extent possible.
René Descartes

GENERIC PROPERTIES OF SPACE VACUUM

As this new millenium kicks off, the sensation of and hope for a renewed light in the course of human history burns even warmer in our hearts. Whether you love the thrill of star gazing as a hobbyist or a scientist, or you are fervently expecting the glorious come back of the God from the heavens, the emotion bears the same name: the fervor of mankind's promotion in the universe. This series of books strives to significantly contribute to this intractable overture of cosmic consciousness towards the Space era. They set forth a novel vision of the universe through a deeper understanding of Space, namely its properties and phenotypes. Indeed, the imminent elucidation of all philosophical, religious and scientific paradigms, which stands for the gateway to the next era, requires a profound revolution in the way we formally and informally visualize ourselves, the objects that we manipulate in the world around us along with the universe beyond us.

The specific objectives pursued in this exposition is in essence to set forth a novel model capable of undertaking a successful interpretation of space deployments in the universe, from microcosm to macrocosm, and to subsequently put it to task. In this first cast of our work, said model will specifically and fruitfully assist us in undertaking the long-sought theoretical derivation of the fundamental physical constants, a paramount goal of modern physics. One may readily understand the importance of that objective, given that the physical constants represent the structural pillars of all of physical matter. In subsequent volumes of this publication, our model will help shed light on an astonishing array of matters germane to Biology and the Human Sciences, beyond the strict physical order, attesting thereby to the true signature of the long-awaited *Theory of Everything.*. This rendez-vous with space is one for the layman, better accustomed to common sense interpretations of events in everyday life, as well as for the educated and well-rounded mind, better trained to elaborate interpretations of life on an individual, collective and perhaps universal level. So much far-reaching is the space model because space traverses all levels of expression of matter and life. Let us further take a more detailed look at this model.

1.1. Mass vs. Space

Gravitation, the force of attraction between two macroscopic objects, has tricked our common sense into believing that the only way things come to existence is through the property of weight. We find ourselves naturally tied to the surface of planet Earth, along with all other objects around us subject to the same condition. The interaction between these many earth-bound entities produces the events and phenomena that are meaningful to us. Thus, to common sense the property of weight defines the existence of any one object: if some *thing* can hit or be hit, we take it as an existing object and vice versa. However, what qualifies all objects around us to cause an impact or be impacted upon, is rated about 10 times less than what common sense tends to make us believe. In effect, according to Newtonian physics, we must divide the weight of an object by 10 approximately in order to circumscribe the real quantity that makes it subject to impact or collision.

Sir Isaac Newton posed that:

$$W_{eight} = m_{ass} \times g_{ravity}$$

In other words, when we see an object fall from the sky, the quantity inscribed within the object that will make it **hit** the ground is in reality approximately 10 times less than what our common sense tends to appreciate, or maybe to fear if the object is about to hit us. This quantity, which is embedded in all objects and qualifies them as such according to Newtonian mechanics, is called their *mass m*. Thus, from a more careful interpretation than common sense appreciation, from a Newtonian viewpoint that is, an object exists inasmuch as it exhibits mass. This definition is mathematically described by manipulating the above equation and converting it into the following:

$$m_{ass} = \frac{W_{eight}}{g_{ravity}}$$

The value of $g_{ravitation}$ is *9.8 m/s²*, that is, almost *10*, if m_{ass} is measured in kilograms. But what is the definition of mass? It is interesting to notice that mass as a primary property of objects has never been defined intrinsically but in relation to other quantities. The truth is, despite all the elaborate developments brought about by Newtonian physics and later by relativistic mechanics, no one to date has ever conceived or proposed an *exact definition* of mass, because no one knows what mass is *per se*.

So far no physics school, neither Newtonian, nor quantum or relativistic has ever proposed an intrinsic definition of mass (the stuff that it consists of), notwithstanding its stature as a fundamental physical quantity. Even though our inner senses dictate that it relates to the faculty of compactness, entirety or indivisibility of an object or the components of thereof, the difficulty to explain explicitly *what* it is still remains whole. As we attempt harder and harder to conceptualize the notion of mass, it becomes interchangeable with the concept of quantity, a truly mathematical concept. There lies indeed the reason why in physics mass is categorized as a *scalar quantity*, in other words, a pure quantity. Paradoxically, the only reason why we know of mass is because there exists something called space or *vacuum*, which we usually disregard. This is to say that mass veritably can only be defined by opposition to space or vacuum: mass is *some thing* while space is *no thing*.

Hence, by relating mass to space it is possible to set forth a formal basis of interpretation of its physical meaning. It is only because of the existence of space that we can have a witful representation of mass. We know of *some thing* because we know it is not *nothing*. The inverse is as much true. Space or vacuum is truly the absolute absence of *some thing*. Physicists contend that the problem of definition of mass is a philosophical matter which physics as a scientific discipline should not be concerned with. However, there is much to learn from such a philosophical exercise, since it may indeed yield a new framework of interpretation of physical nature. Clearly, from our perspective, said space or vacuum entity acquires particular significance. Space not only assists us with a more formal interpretation of mass, but it also stands for a crucial component in the makeup of all objects in the universe.

Space or vacuum deployments are omnipresent in the universe, from quarks to solar systems to galaxies… Space is meaningful, however absolute its absence of contents may appear. Our common form of appreciation of objects gives prominence to their property of mass. Nevertheless, their vacuum or space content does play an important role in their physical behavior. This result is one of the most important consequences of relativistic physics. Albert Einstein proposed that in the vicinity of large masses the structure of *spacetime* is curved. From our standpoint, however, there is no such thing as *spacetime* but simply space, and the properties of space are not limited to cosmic space but generalized throughout all of its forms of deployments both in the macrocosm and the microcosm. It remains all the more important to comprehend that *nothingness* or the state of vacuum is endowed with specific properties, no matter the type of space deployment under consideration.

1.2. Space and Time

The concept of time has misled us into misconstruing space. From a Newtonian standpoint, time stands for an absolute and universal reference which can be used to measure all events occurring in the universe. This intuitive concept of time proves to simply be incorrect. If you are moving and observing a moving object, while someone else, being at rest or not, is equally observing the same object, one of you will see the object moving faster than the other person can appreciate. This result means that there is no absolute value for the quantity that we call speed of an object. Because no one is at the center of the universe and every one thing in it is in motion, no one can admittedly claim his or her observed speed to be the proper or correct one. Since speed is mathematically dependent upon one constant and only one variable, which is time, no absolute speed means no absolute time. Therefore in our common way of interpreting nature, which involves the idea of speed, something is at fault. Our intuitive idea of absolute and universal time is what is at fault. If there is no absolute time, is there any time at all?

What's more, no matter how fast you and any other observer are moving, assuming that both of you could measure the speed of a travel pulse of light, you would both see it move at exactly the same speed! This second result destroys the idea of time itself in our sense. Were the observers still to appreciate different speeds for the light pulse, the idea of time might then still be valid, since they would each appreciate some speed for the beam, albeit different, like in the above case. The invariance of the speed of the object under observation, independently of the observers' travel speeds, represents a strong rejection of the idea of absolute speed and thus of the implied idea of time. Albert Einstein imagined and proposed an interpretation for defaulting time: the concept of *dilation* of time. He was not ready to ring the bell for it yet. Let us look further into the notion of time and how Albert Einstein dealt with it in constructing the Theory of Relativity.

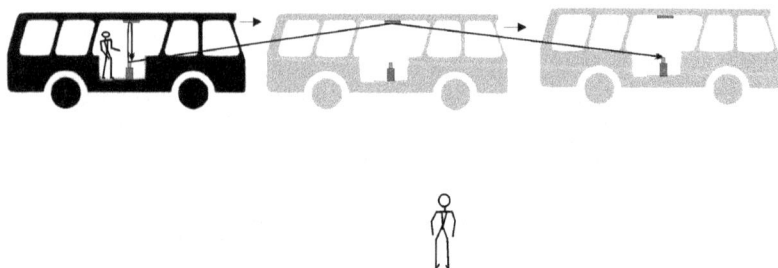

Fig. 1.1 Travel path of a laser beam in a fast-moving vehicle
as seen by two observers

Time basically means for us all *duration*. Suppose that you are in a moving train (Fig. 1.1), and hold in your hand a small laser pen whose beam of light you direct upward to a mirror attached to the ceiling so that the light reflects back into the aperture of the laser device. An amount of time T_1 will have elapsed between the instant the light left the laser and the instant it reaches back the aperture of the laser pen. Now, to anyone standing outside the train, a friend maybe, whether he or she is at rest or not, the beam will appear split in two beams. Due to the movement of the entire train, the external observer will also see two different positions for the pen, one at the time the light left the laser pen and another one at the time it returns to it. The total path of the light will appear as two sides of a triangle to the external observer, not anymore as a straight line like you, the first observer located in the train, duly perceive it. Obviously, this second path will be much longer than the one you observe in the train yourself. Since the speed c of the beam of light remains the same, this also means that the light will take longer to cover that distance. Consequently the total time duration T_2 measured by your friend, the second observer, is larger than the time duration T_1 that you observed in the train for the round trip. This is a contradictory situation where the same event is measured to last different amounts of time or different time lapses. Albert Einstein proposed to conceptualize this result as a dilation of time and showed how to relate T_2 to T_1 with the following relation:

$$T_2 = \frac{T_1}{\sqrt{1 - \left(\dfrac{v}{c}\right)^2}}$$

where v is the relative velocity between the reference frames, and c the speed of light.

Therefore time is also relative. The rate at which time elapses for you depends on the speed at which you are moving. Clearly, from that perspective, every observer may have his own scale of time. While the concept of relative speeds is acceptable because they are made subject to a common reference which is time, the concept of relative time is not quite acceptable unless we can provide an explanation of what these relative time durations are referenced to. We express speed as a rate of change of distance in time in terms of dx/dt. If time is relative, how can we express its rate of change? The idea of rate at which time passes is formally unclear and to further make it depend on the speed of the observers themselves is to create a vicious circle whereby speed depends on time and time depends on speed. Deemed to resolve this serious problem, A. Einstein was led to conceive that space and time are not independent from one another, that they are *entangled*. That is to say, the ultimate implication of Special Relativity is to make the rate dx/dt equal to dt/dx. He referenced time to space. This is implicitly the basis for the relativistic entanglement of space and time, the structure he coined *spacetime*.

From either a Newtonian or a relativistic vantage, if time is not a primary, exclusive and independent state of matter (or a dimension as it is traditionally visualized), the only other state of which it may become a variable is space, not mass. Mass is naturally and definitely best conceived as a pure unredeemable scalar quantity in all forms of mechanics. As previously stated, the mathematical formulation of construing speed as dependent on time and time as dependent on speed is:

$$\frac{dx}{dt} = \frac{dt}{dx}$$

This relation can only be true if the differentials dx and dt can be equated to one another: $dx = dt$. This would translate into equating time and space, or simply ringing the bell for time. A. Einstein was not ready for a mathematical description of the macroscopic universe in complete dismissal of the time variable. Despite the great success of Special and General Theory of Relativity, essentially attributed to their ability to correctly predict various observations in the physical order, the Theory remained of difficult application at the quantum level because time or time-driven dynamics did not seem to occur in that realm, much to A. Einstein dislike. He proposed in that regard an experiment that could not be carried out at the time. To our good fortune, the developments in laser-optics technology in the 1980's made it possible to realize in the laboratory a good approximation to the original concept (the EPR Paradox). Take two identical units of the smallest particle of light, two photons, and allow them to travel in opposite directions to each other in deep vacuum, can anyone influence by any means the other? Let us first remember that the epitom of Relativity is the invariance of the speed of light and that this speed c is the maximum possibly attainable by any physical object. Since each particle is moving away from each other at speed c, the maximum attainable, any interaction from one particle toward the other must travel at a higher speed than c in order to reach its mate and bring any information to it. A hypothesis held impossible by one of the fundamental postulates of Relativity. However, it has been observed over and over for now three decades that both particles always respond in the same way to individual random events affecting only one of them, as if they could indeed inform one another in some unknown fashion.

Other variants of this experiment have been realized, notably the one by physicist Raymond Chiao, all of which always end up casting some faster-than-light result. Physicists have been somewhat uncomfortable with these results because they suggest a violation of the cornerstone rule of General Relativity. Instead of retrocausal activity in quantum physics as increasingly hypothesized of late, could there be non-observable *locality* beyond 3-dimensional space where faster-than-light speed is possible and can in fact produce *effects* that are observable in 3-dimensional space? Could the Theory of Relativity as a whole be no more than a particular case of a more general physics theory from which it can be deduced when the speeds under treatment are neighboring the speed of light? A somewhat similar case to Newtonian physics which can be deduced from a relativistic treatment when the speeds involved are only a small fraction of the speed of light.

Those hypotheses would suggest the death of time as we know it. The dismissal of the time variable is well established in pure Quantum Mechanics, the study of events in the atomic and to some extent the sub-atomic world. Because of lack of understanding of the larger implications of Quantum Mechanics, quantum physicists have not been able to settle those debates around the EPR paradox satisfactorily. Furthermore Quantum Mechanics does not include a treatment of gravitation, a major effect generalized at cosmic scales, in what has been considered a significant problem. Despite its atemporal approach, which particularly suits the effect described above, Quantum Mechanics is unable to predict the gravitational behavior of celestial bodies. It is interesting to note that the quest for a unified theory, with the capability to predict the universe as a whole and in all of its details, has suggested the eventual fusion of time and space. Effectively, in Richard Feynman's incipient proposal to formulate Quantum Mechanics in terms of the integral of all possible trajectories, because we would need to express time with imaginary numbers instead of real numbers (earliest version of retrocausal interpretation to which we do not subscribe however), one may arguably conclude that the historic distinction between time and space totally vanishes and ultimately time is reduced to space for all practical purposes.

1.3. Properties and Faculties of Space and Mass

Let us convince ourselves: *there is no time anywhere but space everywhere.* We have stipulated earlier that space can only be defined in opposition to mass and vice versa. There is no prior principle from which an ontological definition of mass and space can be derived. There is no absolute physical or conceptual separation between them. From our standpoint, they are prime values non-deducible from any previous form of reality and they relate to one another in exclusion of any other possible form of reality. Nothing else can affect space and mass, from nowhere do they come. Mass, call it a *thing*, and space, call it *nothingness*, are the basis of all forms of objects as well as events in our reality.

In studying the microcosm, physicists had pinpointed the existence of small particles of matter, neutrons, protons, electrons, ... which they originally believed to be small quantities of mass. Soon they discovered that these particles could exhibit properties not attributable to their mass. This discovery contributed to the development of quantum physics through the work of Louis Victor de Broglie, Erwin Schrödinger and many other brilliant minds. In particular, some time after Max Panck's discovery of quantization of energy, particles were experimentally found endowed with wave properties. Since the particle was visualized as pure mass, or a pure scalar, its capability to harbor a wave function came as a surprise and somewhat of a conundrum. In effect something different from the mass must exist within the particle to produce the wave behavior. To date quantum physicists do no hold an all clear physical representation of the wave function embodied in particles, notwithstanding the adequacy of the related mathematical developments. The interpretation of the physical meaning of involved functions, operators and equations do not come as an easy task. Hence physicists have resorted to many different interpretations of these results which are often at odds with each other. The truth is that the wave function contained in particles is an expression of their space component, simply put the *vacuum* or *void* that is a part of their ontology.

Going beyond the laws of gravitation exposed by Isaac Newton, A. Einstein developed a mathematical basis through Relativity to explain gravitational effects occurring in the macrocosm. He concluded that gravitation is a geometric effect released by the inner structure of *spacetime*, say just space. According to this view, when a planet for instance orbits the sun, it does so not because of a force of interaction between the sun and the specific planet, but because the latter follows a particular space path that is curved. From Newtonian mechanics, the closer two bodies are to each other, the stronger the force of gravitation (or attraction) between them. Consequently, when a planet orbiting the sun arrives at its closest point to the star it should never be allowed to move away from it. Nevertheless, this is the case with Mercury which, instead of following its elliptical path as it gets closer to the sun, surprisingly diverges from the trajectory, describing a divergent arc of a circle around the sun. A postulate of relativistic treatment is that around large masses space is curved. Since the planet follows the inner structure of the space tapestry, it can only diverge in close vicinity of the sun. Therefore space in the macrocosm is endowed with geometric properties.

Let us now generalize by establishing that geometry and motility are two primal properties of space at all scales of the universe. Things do not move because of their mass. They move because of their space component. Movement is ultimately transmitted from one body to another not through mass but through space. When something is moving, it is the space distributed within the object that produces the movement. How can that be possible since space is *nothingness* and infinitely *dilatable*? Because space is nothingness, in a sense it does not make much of a difference in terms of its fabric whether it is stretched or shrunk, it remains nothing. Conversely, if something can be stretched or shrunk and remain same, this *thing* must be nothingness, because nothing we know of in the universe, from our usual quantum or mass point of view, can demonstrate these qualities. This thought exercise is far from trivial and is key to an understanding of the structure of space. The following is a law: *space is infinitely elastic*. It is so infinitely elastic that, no matter its degree of propagation, it always appears to us as nothing and has never deserved our slightest attention. The dilation of space means that it is in propagation. From our standpoint, those two terms are interchangeable. Space is always in propagation, it is indeed a *propagation field*, every where it appears, from microcosm to macrocosm. One way to look at any physical object is to consider it a population of discrete masses or scalar elements (dense quanta) articulated to a propagation space field which they follow because they are subject to it. **In its purest state** (an unreachable asymptotic limit)**, the degree of propagation of space would be absolute and would give rise to instantaneousness**. In physical reality however, where it is coupled to scalar elements, it loses innate dynamism and develops distinct degrees of propagation, as we shall see later. We will give mathematical support to these concepts further down in this exposé.

From a quantum-mass standpoint, the universe is populated at all scales with individual objects, distinct and segregated from one another. From a space vantage, all objects are directly related to one another, not by juxtaposition, but through immersion or superposition. Recall that space is infinite and so the particular propagation field of an object expands to infinity. Consequently, no matter the distance between two objects, they are directly related because they are embedded in one another through space. Fourier analysis has instructed us that there is no limit to the superposition of waves. It tells us that what appears to be one particular wave of a specific frequency may be treated as an infinite number of specific harmonics. Fourier treatment is not just a theoretical model to perform electrical analysis, it reflects the exact behavior of space, and only because space is the essence of the wave phenomenon does the Fourier treatment become possible. In other words, when we see one "form of space" or one thing with wave characteristics, it is in fact the ultimate sum of an infinite number of space forms which bear a definite symmetric relationship between one another. We will further elaborate on this particular topic later.

This naturally brings us to consider the second property of space: *symmetry*. Space is not only a propagation field, it also holds latent states of symmetry that become apparent as geometric forms, or just forms, if you will. There is no objecting to space as the origin of the dimensioning of physical objects. Space veritably provides the form or shape, simple or complex, extrinsic to the dimensioning of objects. It is an inherent property of the space field which organizes the geometry of the object. This property of space which becomes apparent in the geometry of objects is what we call state of symmetry. A more practical approach for grappling this very abstract concept is probably to start from the end toward the beginning. Consider this: all forms are forms of symmetry. Whether an object exhibits a very sophisticated sinuous shape or simply a linear or planar shape, it is eliciting symmetry. It is easy to understand that a circular or spheroid shape can be broken down into parts that are all similar to each other. These fractional parts can be superimposed on each other in perfect match, they are interchangeable. The most complex sinuous shapes can be broken down into circular, elliptic, parabolic... fractional parts that are interchangeable with each other, whereas all of the parts from a linear or planar shape are almost absolutely interchangeable. Therefore, all forms are forms of symmetry. Now, if by some magical means, one could take out of an object and its components all of their discrete masses, what would be left behind? You would hypothetically leave behind a space field, which, no matter how deeply intangible, did articulate the masses into the particular geometry or form of the object. Hence, it would hypothetically remain with this specific potentiality of geometrization. This potentiality, simple or complex, is a state of symmetry. Much of what is called a wave function in Quantum Mechanics is attributable to it. By simple observation, you would never know of the geometric potentiality of this space field, if conserved, until it gets involved with a re-distribution of discrete masses. It will then re-inform them according to its specific inner geometric orientation or qualities.

1.4. The Spatial Nature of Waves

Our notion of waves stems from the observation of oscillatory motion. We have learned to use trigonometry to give an exact account of this type of motion. Consider a point on a circle and let it travel eternally on the perimeter of the circle (Fig. 1.2). The projection of this point on the x-axis of the trigonometric referential describes in time an oscillatory motion, travelling endlessly from one end of the axis to the other. Let us now presume that you lay an endless stripe of paper through the x-axis perpendicularly to the plane of the circle, and let the paper move at constant speed. If the projections of the circling point could be set to cause a manner of dots on the paper, we would quickly start to notice a continuous sinuous form on it, which our mathematics designates the eternal sinusoid. All that we know of oscillatory motions is captured in this sinusoid and the sinusoid describes the sum of our knowledge of wave motion. We visualize the eternal sinusoid in time by laying it down on a coordinate system where the horizontal axis stands for the time variable t and the vertical axis the function of time $f(t)$. We will now briefly show how to arrive at the usual time expression of the eternal sinusoid.

The equation describing the position of an object traveling in uniform rectilinear motion is:

$$x(t) = v_0 t + x_0$$

where $x(t)$ is the travel distance at any time, v_0 the linear velocity of the object at time zero, the initial time, and x_0 the distance already traveled at time zero, if any.

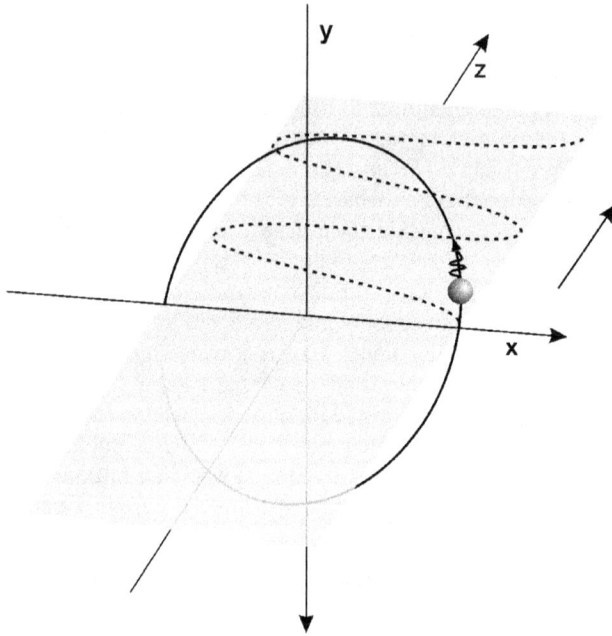

Fig. 1.2 Eternal sinusoid resulting from projection of a circling point

Looking back at our circling point, we can express in the same fashion the distance it travels around the circle as:

$$\theta(t) = \omega_0 t + \theta_0$$

where $\theta(t)$ is the travel circular distance at any time expressed in radian, ω_0 the angular velocity of the point at time zero and θ_0 the distance already travelled at time zero, if any, and expressed in radian.

Both equations are seemingly identical. The only difference concerns linear quantities in the first one as opposed to angular quantities in the second one. We know, according to the trigonometric law of sine, how to express an angle:

$$\sin \theta = \frac{opposite\_side}{hypothenus} = \frac{y}{x}$$

When θ is expressed in degrees it means the angle formed by the point on the circle, the origin of the referential and the right-side portion of the x-axis. When θ is expressed in radian it simply means the arc subtended by the angle, which is the angular distance whose time expression we are looking for. Whether the angle is expressed in degree or radian, the law of sine holds always true. We can rewrite the last equation in the form:

$$y = x \sin \theta(t)$$

and by replacing $\theta(t)$ by the value we have found earlier, we obtain:

$$y = x \sin(\omega_0 t + \theta_0)$$

We replace y by $f(t)$ because the expression on the right side is a function of time. In addition, we can easily see that x is the maximum value that the function can reach: since the maximum value of a sine is 1, the maximum value that the right side of the equation can reach is x itself. For that matter x is usually called the amplitude of the function, it is preferentially designated by A and takes the unit associated with the nature of the phenomenon, distance, current, voltage, flux, etc. The final general expression of the oscillatory phenomenon is then:

$$f(t) = A_m \sin(\omega_0 t + \theta_0)$$

This expression describes and gathers all we know about the physical nature of wave phenomena in general. The details of its derivation have been reported here in order to set a baseline for the discussion. There is no denying that all wave phenomena involve motion and repetition. Repetition means symmetry or interchangeability. The period of a wave is defined as the time it takes to complete itself before the cycle starts over. The half-period itself is a second element of symmetry. Any set of two consecutive periods of the eternal sinusoid are absolutely interchangeable, and within the period there is odd symmetry (symmetry with respect to the axis-crossing point) between the two half-cycles. Our usual mathematical expression for wave phenomena encloses indeed an account for symmetry. The odd symmetry between the half cycles springs from the even symmetry between the half-circles of the trigonometric circle itself, which stands for the roots of the model; the symmetry between the periods for themselves derives directly from the trigonometric circle as a whole. Therefore the forms of wave phenomena we know how to model mathematically are those enclosing circular/spherical symmetry.

The ordinary implicit assumption that the nature of all wave functions is always the same is totally uncertain. We used to believe that the circular model set forth in trigonometry is general and applicable to all forms of oscillatory phenomena. Until we discovered that there exist certain objects in the universe that demonstrate oscillatory properties but do not demonstrate oscillatory motion like the pendulum or electrical current in a wire, for instance. The discovery that elementary particles moving in straight line could also exhibit oscillatory properties, i.e. interference and diffraction, as do conventional wave phenomena, posed a barrier to our understanding of the atomic and sub-atomic structure of the universe. Despite a fierce intellectual battle conducted through the use of differential calculus and operator mathematics, the wave function associated with particles is still not clearly understood. In addition, Quantum Mechanics can only explain the organization of the hydrogen atom, the simplest of all. It is important to understand that a wave phenomenon is not only a form of motion but also a form of symmetry. The trigonometric model principally takes into account the mobility aspect of the phenomenon. The symmetry facet does need to be accounted for as well. To do so, we may need to replace the trigonometric circle by other forms of symmetry and derive an expression for each form of symmetry. But we would first need to study the different orders of symmetry prevailing in the universe, the different inner states of symmetry that space may be endowed with, that is. This will become the subject of later treatment.

1.5. Wave functions in Quantum Mechanics and Relativity

The real contribution of Quantum Mechanics to physical science is arguably not to have modeled the atom, a microscopic object, basically as a wave function, however important this form of representation proved to be. The success of Erwin Schrödinger's treatment is mostly due to the set of conditions that he posed for the wave function to give an adequate account of what the atom appeared to be: a particle limited to move freely in a box. He started with the travel electromagnetic wave equation, boldly defined a wavelength λ for the particle which he incorporated to the equation turning it into:

$$\theta(x,t) = A_m \sin\left(\frac{x}{\lambda} - vt\right)$$

This equation depends on two different variables: x, the distance travelled by the wave (or the particle) and t, the time variable. Since we introduce a second dimension in the wave phenomena, we must ensure that the variations of the wave in this dimension are also regular or "symmetrical", otherwise the motion would no longer be a wave motion. We express this condition mathematically by stipulating a specific condition between the rate of change of the wave $\theta(x, t)$ with respect to the time variable t and the rate of change of the wave with respect to the distance x. The ratio of these second-order partial derivatives must be equal to $(\lambda f)^2$:

$$\frac{d^2\theta}{dt^2} \cdot \frac{d^2\theta}{dx^2} = (\lambda f)^2$$

Or, in proper partial derivative notation:

$$\frac{\partial^2\theta}{\partial t^2} \cdot \frac{\partial^2\theta}{\partial x^2} = (\lambda f)^2$$

This condition is of no surprise should we remember that, in differential calculus, the process of differentiation affords the study of concavity, in fact the curvature or curvilinear evolution of functions. If t were a tangible dimension, like z in a three-dimensional frame of reference, this condition ensures that the motion would then describe the eternal sinusoid in three dimensions. Imagine a plane folding in sinusoids.

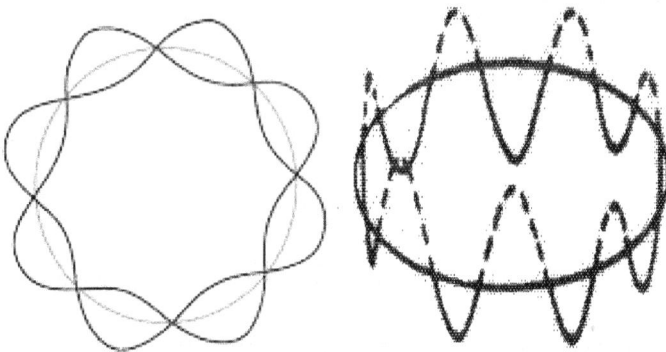

Fig. 1.3 Virtual depiction of a stationary wave from two different angles

E. Schrödinger further established that this wave associated to a particle limited to move within a box must be stationary. Otherwise this associated wave would end up fading away. If the walls of this imaginary box were tangible, this means that the supposed sinusoids described by the particle ought to hit the wall only in the node phases of the sinusoids. Otherwise the reflected waves would be out of phase with the incident ones, and the wave phenomena would be quickly absorbed by the intangible walls. However, these walls bear in fact no physical ability to be impacted by the particle, because there is no real box, so that the wave must yield the stationary effect all by itself (Fig. 1.3). Consequently, the original wave must be a composite one: two waves of the same frequency (same sinusoid) but with different phase. The expression of this composite wave is:

$$\theta(x,t) = A_m \sin 2\pi\left(\frac{x}{\lambda} - vt\right) + A_m \sin 2\pi\left(\frac{x}{\lambda} + vt\right)$$

$$= 2A_m \sin 2\pi\left(\frac{x}{\lambda}\right)\cos 2\pi ft$$

Because of the ad hoc introduction of the terms of this specific composite function, we ought to make sure that it obeys the conditions for wave behavior. When we apply the above differential condition, that is to say taking the ratio of the second-order derivative and making it equal to $(\lambda f)^2$, surprisingly the time component of the composite wave (cos $2\pi ft$) cancels out and the condition becomes:

$$\frac{d^2(2A^m \sin(2\pi x / \lambda))}{dt^2} = \left(\frac{2\pi}{\lambda}\right)^2 2A_m \sin\left(\frac{2\pi x}{\lambda}\right)$$

Let us make $2A_m sin(2\pi x/\lambda) = \Psi(x)$ giving the function a denomination. The above condition then becomes:

$$\frac{d^2\Psi(x)}{dt^2} = \left(\frac{2\pi}{\lambda}\right)^2 \bullet \Psi(x)$$

Recapitulating, in order for our wave-particle to act as limited within an intangible contour, it must behave like a stationary wave such that the second derivative of this wave function is equal to the wave itself, except for the multiplication by a constant term. If we remember that the first derivative of a sine function is a cosine function and the derivative of the latter a sine function again, we see that the sine function satisfies the posed differential condition for the composite wave. As to the argument of this function, it must be such that when we differentiate it we obtain the coefficient $(2\pi/\lambda)^2$ present in the right-hand side of the equation. Consequently the wave function must be:

$$\Psi(x) = A\sin\left(\frac{2\pi x}{\lambda}\right)$$

We do not really know what this wave expression physically means. We know that the particle does not certainly have a classical wave motion, akin to a sinusoidal perturbation moving along a straight rope. The latter is the classical case of a motion as a function of time, whereas Ψ is only a function of x. The argument of the function $(2\pi x/\lambda)$ does not lend itself to a clear interpretation either. What is x/λ? This wave is what is called a De Broglie wave. Louis Victor De Broglie established that to every movement is associated a wave which pilots the movement, no matter how big or small the travel object is. In fact, the argument $2\pi x/\lambda$ is an equivalent of the argument $w_0 t$ in the general expression of the sinusoid. We know that:

$$\omega_0 t = 2\pi f t$$

where f is the frequency of the wave or the number of complete circular cycles per second. Since the frequency is the inverse of the period T: $f = 1/T$,

$$\omega_0 t = \left(\frac{2\pi}{T}\right) t$$

One way to make sense out of the De Broglie wave function is to compare it to the general equation of wave motion. That is to equate the arguments:

$$\left(\frac{2\pi}{T}\right) t = \left(\frac{2\pi}{T}\right) x$$

The role played by time is now played by distance or *space*. The variable t may be assimilated to x and the period in time T assimilated to the period in distance or the wavelength λ. That is to say, the movement of the object is not rated to change in time anymore but in space. In other words, the continuous dilation of space produces the movement. Moreover, the cancellation of the time variable is a confirmation of this interpretation. This conclusion is similar to the one drawn from our critical study of the relativistic treatment. Furthermore, the sine expression tells us that the De Broglie wave of the constrained particle possesses circular/spherical symmetry. The translation is that the structure of the space confines where the particle is constrained to remain possesses spherical symmetry. This form of symmetry will become apparent in the coupling of the particle with this extension of the space layout. From our perspective, the particle is not merely somewhere, the space location where the particle is moving is indeed an intrinsic part of the particle. What's more, it is this space location or field which is "piloting" the movement of the particle through its inner process of dilation. Its inner property of circular symmetry will lend to the distribution of the particle the apparent geometry of a sphere. The object is no longer a particle, it is a space sphere with a "distributive" particle.

From a quantum-mass perspective, the reality of the particle restricted to a space location is a kinematic event, from a space perspective it is rather an object because the particle is *coupled* to the immediate space deployment surrounding it. You no longer observe the particle as an object localized somewhere, you observe a particle describing a spherical geometry, a new object. This object is truly an *observable* no matter how close to the speed of light the speed of the inner moving particle is. We call this object a distributive object. It is endowed not only with proper space but also with proper mass. The mass is not localized, it is distributed. If the particle were merely somewhere, no matter how limited the extent of its motion, its mass would be preserved because it is not radiating energy: its pilot wave is stationary. But because the mass of the particle is redistributed throughout its immediate space deployment in what we call its coupling to space proper, the new object is born with less mass than that of the discrete particle itself, its exclusive mass contributor. In other words, the new mass loses localization and as such scalar magnitude. This effect is known as a mass defect in Quantum Mechanics.

1.6. Physical Space as Active Void

While the notion that pure vacuum is not an absolute void might still remain quite foreign to the larger public, it has been floating in Quantum Electrodynamics since the early 1930's. This line from Joseph Silk is often cited to explain the representation that the field holds of pure vacuum:

The quantum theory asserts that a vacuum, even the most perfect of vacuum devoid of any matter, is not really empty. Rather the quantum vacuum can be depicted as a sea of continuously appearing and disappearing particles that manifest themselves in the apparent jostling of particles that is quite distinct from their thermal motions... At any given instant, the vacuum is full of such virtual pairs, which leave their signatures behind, by affecting the energy levels of atoms.

Others might equally add fleeting electromagnetic waves to this constant annihilation of particle pairs or pairs of artifacts inside of pure vacuum. While in Quanto-Geometry we may subscribe to this view, our primary intent however is not to describe what there is beyond physical space vacuum and how that affects the latter, but to describe its own seminal structure as it appears in front of us. Quanto-Geometry asserts a whole order of things in pure vacuum or nothingness that amounts to a structural order in its fullness such as:

- Symmetric latencies

- Geometric (Morphological) latencies

- De-ploy (Un-Fold) and splitting capabilities

- Active propagation

- Parameterized entropy and Shell configuration

We will further elaborate along those lines as we develop this proposal. We expressly caution that Quanto-Geometry should not be assimilated to a Quantum Field Theory, notwithstanding this exposition being substantially run in contrast to the latter. The distinction between these two views of matter essentially rests on the nature of abstractions and analytical framework at work in both theories, despite their addressing the same objects, as we shall see in the next chapters. We believe that comparatively Quanto-Geometry presents a much wider analytical scope as much as better targeting and, more significantly, effective results in the task of responding to the string of paramount yet unresolved questions in the physical sciences.

1.7. Conclusion

In this Chapter, we have examined the general properties of space vacuum. We have revisited Special Relativity and Quantum Mechanics to show their limits and their implication in relation with this new vision of space in the universe. We do not wish to overwhelm the uninitiated reader with more mathematical developments than necessary and will tend to emphasize the interpretative aspect of the exposition. All too often, the desire to develop supporting mathematical formalism in order to secure accuracy, all loadable though it is, supersedes the abilities for assertive and sensible interpretation of results cast by the formalism in order. If the late paradigms are to be overcome, physics and physicists ought to reconcile and engage with philosophy, and creatively tackle the difficult issue of interpretation of reality from the fullness of the formalism that has now been known for about a century. The biggest problem in the field in our view has been and remains the inability to overcome the notion of time as a consequence of the inability to properly envision the inherent dynamics of space vacuum, despite unrelenting attempts. The final assault toward Grander Unification requires many bridges to be cast and many doors to be knocked open.

Chapter 2

*God used beautiful mathematics
in creating the world.*
Paul M. Dirac

UNIVERSAL MASS-SPACE ARTICULATION
IN QUANTO-GEOMETRY

This journey through space brings us now to the specifics of mass-space coupling. Having established the primacy of mass and space in the makeup of all objects of the universe, along with the general properties of space, our next step is to study the different modes of articulation between mass and space within physical objects making up the universe. At the end of the previous chapter, we have shown how both factors arrive at a mutual articulation. We will now look into the principles that govern this interaction. However, we will first examine the use of Operators in modern physics.

2.1. Mathematical Operators

Everybody has long familiar with simple arithmetic. Since our early schooling days we learned how to perform simple arithmetic operations involving addition, subtraction, multiplication and division as applied to any set of numbers. So for instance, when we see the + sign between two numbers, we know that refers to the operation of adding one of them to the other. Hence the + sign, as all the other arithmetic signs, means a specific type of *operation* to be applied to two or more quantities. We may generalize this notion by stating that an *operator* sign is a transformation index. It instructs us about a specific transformation to apply to any quantity or set of quantities. An operator is thus in essence a transform. Take the expression:

$$9 = 6 + 3 \quad (2\text{-}1)$$

We may transform expression *(2-1)* by multiplying both sides by 2, for instance. We then obtain:

$$2 \times 9 = 2 \times (6 + 3) \quad (2\text{-}2)$$

$$18 = 12 + 6$$

Essentially, the equality remains true but we have transformed it by applying on both terms the exact same operation, which is literally 2 x. Therefore, 2 x as a whole is an operator. It means: do the operation of multiplying the terms of the equality by two. So we are applying here our operator to not only other numbers but numbers that are already involved in a specific relationship, in this case a whole equation. We may avail ourselves of the convenience to rewrite equality *(2-2)* in a more concise format as:

$$2 \times [9 = (6 + 3)] \quad (2\text{-}3)$$

Here, the expression inside the brackets is a simple arithmetic expression, it might have been a function instead however. Then it would become a specific case within a higher level of abstraction. It might be, among many other alternatives, an instance of the function:

$$f(x) = 2x + 3$$

where x might stand for the number of lies commonly detected in the testimony of a certain category of criminals, in a court trial! The function means to multiply by 2 the number of lies found and add 3 to that result, in order to obtain the total amount of lies $f(x)$ in a testimony. In the instance that you detect 3 lies, the expression becomes $f(x) = 6 + 3$. In that case, it turns out to be literally equivalent to expression *(2-3)* and reads:

$$2 \times [f(x)] \quad (2\text{-}4)$$

Giving to 2 x the standard operator notation, expression (4) reads:

$$< 2 \ x > [f(x)]$$

Our operator $< 2 \ x >$ is the simplest of operators, merely meaning to multiply the function by 2. We might want to perform that operation in order to obtain information about the age of the criminal. Due to its simplicity, this operation is probably best accounted for inside the function itself, or otherwise by simple arithmetic. However, we might have an operator prescribing to take the first-order, second-order or fourth-order derivative of the function rather and then to multiply it by 2 so that it reads $< 2 \cdot d^4/dx^4 >$. We might want to do so to extract pertinent information about the age of the criminal.

Operators are useful in modeling complex systems where the functional representation is unable by itself to cast proper solutions to the unknown variables under scrutiny. The use of a specific operator allows us to test for a specific pre-conceived behavior of the system. One very widely used operator in Quantum Mechanics is the Hamiltonian Operator. It is an operator of kinetic and potential energy and stands for the only path toward resolution of quantum systems's variables. Thus operators are specific types of mathematical operations that we generally apply upon complex functions as a first step toward their resolution. If the function reveals itself proper to the operator, then we may resolve the variables through the *eigenvalues* of the operator.

Above all, it is important to understand that an operator is a pre-conceived abstraction about the behavior of a system: it is a model. Our contention in Quanto-Geometry is that operators are typed, on the one hand, and that, on the other hand, nine fundamental types of operators exist in the physical universe. Therefore, we may acquire ample knowledge about all properties of a physical system because we know a set of fundamental patterns of *operation* - which we shall discuss shortly - according to which the system must behave.

2.2. Quanto-Geometric Coupling

The death of time ultimately signifies that all events in the universe may be observed as objects and all objects as an articulation between mass and space, regardless of the phase of matter under observation. The analysis conducted in the previous chapter asserts that the implications of the Special Theory of Relativity are equivalent to those induced from Quantum Mechanics. This strongly suggests that there is no crucial difference between the laws governing the microcosm and those ruling the macrocosm. This general law exists and its name is *Quanto-Geometry.*

Since all objects enclose space, their mass always qualifies as *distributive* scalar, even though in some conditions the mass appears to be localized. We will then gradually abandon the term mass formally used to describe the tangible *thing* that makes up all of universal objects in favor of the term quantum. The quantum notion provides indeed a more general framing of the reality of mass in an object. On another hand, we have seen that the coupling of mass to space gives birth to *geometry*, that is to say the observable form of objects. Hence the term Quanto-Geometry which signifies the articulation of mass and space in an object. Quanto-Geometry has a formal mathematical expression that we shall analyze later on in this Chapter. Visualized in its purest form, space displays an infinite degree of inner propagation. Visualizing mass in its purest form reduces it to an absolute level of concentration or localization likening it to a pure scalar. Nevertheless, none of these absolute forms of mass and space exist in the universe because all masses are coupled to space and all apparent particular extension of space is coupled to at least one discrete mass or quantum packet. The articulation of mass to space means a constraint exerted on space by mass and, concurrently, a constraint exerted on mass by space. The constraint undergone by space translates into a restriction in its degree of propagation which would be absolute without it. This constraint diminishes the extensiveness of a *space spread*[1] and brings it to some level of *tension*, expressible down to the form of standing behavior. In other words, articulated space is compressed. The constraint undergone by mass for itself means that it becomes *distributive*, albeit not discretely distributed, for it must follow the integral of the trajectories made available by its associated space extension. This constraint renders mass less localized, turning it into a quantum or dense scalar. This stands for a mandatory implication of the Quanto-Geometric articulation. The principle of Quanto-Geometry represents the most primal of all laws of conservation. It is a more general form of the law of conservation as visualized in Newtonian physics and as subsequently utilized in all other forms of physical theory. To render a prime account of an object is to specify its quantum-scalar degree and the degree of spread and propagation of its space partner: the specification must thus render a Quanto-Geometric account of the object. This Quanto-Geometric portrait is the primary modality of expression of the existence of an object. We call this foremost Quanto-Geometric principle the *generic Momentum* of an object and express it as follows:

$$q \times s = 1$$

where q is the quantum component of the object and s is its space component. The above formula postulates that the product of the scalar component q of any element, abstract or real, tangible or intangible, by the space (or nothingness) component s is equal to unity, *1* standing for the whole of the element.

The above expression derives into a series of forms of articulation that are best captured in a functional representation of the Quanto-Geometric primal principle. Before we delve into that functional analysis, let us first revisit the use of operators in Quantum Mechanics.

2.3. The Hamiltonian Operator in Quantum Mechanics

In the preceding chapter, we have seen that the last form of the condition imposed on the wave function in order to consider it emulating limitation in space is:

---

[1] In Chapter 7 we give a formal treatment to the notion of *space spread*, which is not exactly equivalent to the common notion of *field* in physics.

$$\frac{d^2\psi(x)}{dx^2} = \left(\frac{2\pi}{\lambda}\right)^2 \times \psi(x)$$

E. Schrödinger incorporated to this condition, the proposal by Louis Victor de Broglie that all forms of movements are piloted by a wave whose wavelength *is* $\lambda = h/p$, *h* being Planck's constant and *p* the momentum of the particle. He did as much with Dirac's constant $\hbar$ which is equal to $h/2\pi$. By substituting *h* and λ in the differential condition, it becomes:

$$\frac{d^2\psi(x)}{dx^2} = \frac{2m(E-V)^2}{\hbar^2} \times \psi(x)$$

Further manipulation yields:

$$\frac{\hbar^2}{2m} \cdot \frac{d^2\psi(x)}{dx^2} + V\psi(x) = E\psi(x)$$

where *E* is total mechanical energy of the particle (kinetic + potential), and *V* the potential energy.

The question posed by this differential equation is: what is the wave function which, when differentiated twice and multiplied by $(-\hbar^2/2m)$ and added to the potential energy of the particle yields the same wave function back but now multiplied by its total mechanical energy? When you find this function, you have found the wave function that the constrained particle actually exhibits. Another way of formulating this question mathematically is:

$$\left[\frac{\hbar^2}{2m} \cdot \frac{d^2}{dx^2} + V\right]\psi(x) = E\psi(x) \quad (2\text{-}4)$$

Each of the terms within the bracket means a specific operation to perform on the wave function $\Psi(x)$ which is not limited, however, to be an arithmetic form of operation. The first term instructs us to differentiate and multiply by a constant, the second one to only multiply by V. Each one of these terms stands for an operator because they do a particular operation on the wave function. The first one is related to the kinetic energy of the particle (the constant term is derived in fact from the kinetic expression: $E = mv^2/2$) and the second one expresses clearly and exclusively the potential energy of the particle.

Therefore the first one is designated the kinetic energy Operator $E_k\hat{}$ and the second one the potential energy Operator $V\hat{}$. Their sum $E_k\hat{} + V\hat{}$, a non-arithmetic operation, is designated the Hamiltonian Operator $H\hat{}$. That is:

$$\left[\frac{\hbar^2}{2m} \cdot \frac{d^2}{dx^2} + V \right] = H$$

So that:

$$\mathbf{H}\,\Psi(x) = \mathbf{E}\,\Psi(x)$$

It is remarkable that so far the original question regarding the nature of the wave function which satisfies the confining behavior of the particle is still not answered and, in the aggregate, will not be answered directly. Concisely stated, the answer is: *"well, I don't know what this wave is like, but since the wave originates from a particle with mass and momentum, I know something that will provide further knowledge of some of the dynamic variables of this wave-like particle. I know an operator, the Hamiltonian operator, which is capable of interrogating it successfully."* Despite the systematic method developed by Erwin Schrödinger to obtain the operator corresponding to a specific dynamic variable in any system of particles, it remains universally accepted that Quantum Mechanics cannot bring a direct and precise answer to all of the unknowns of a dynamic system. Quantum mechanics even postulates that all of the variables of a given system cannot be determined with absolute precision but only statistically. It is fair to say that this science is more interpretative than predictive, notwithstanding its success in modeling adequately a number of known physical and chemical effects at the atomic and molecular scale.

2.4. The Quanto-Geometric Principle

In interpreting the quantum Hamiltonian operator, we may appropriately liken it to a lens that allows the observer to screen the kinematics of atomic elements. In order to render their observation possible, this screening lens must be transparent to their inner structure. Put differently, this model must *mold* the atom close enough in order to yield an adequate account of its structure and its subsequent kinematics. In summary, the Hamiltonian operator proposes a form of bi-structure of atomic objects and correlated phenomena. Since it is a conjunction of two other operators, it so proposes to screen every atomic event by considering their kinetic and potential energy implications on the basis of a specific mathematical scheme. The order of application of each operator to the unknown is inconsequential, since the expression of the operator dramatically excludes the time variable. In that regard, it makes no difference whether you use on the unknown the kinetic operator first and then the potential operator or conversely. The result will remain invariantly the same in either case. This makes of the Hamiltonian treatment at the outset a structural analysis rather than a historic or kinematic analysis.

The principle embedded in the Hamiltonian is valid not only for the study of the atomic world but for all structures present in the universe. It is a particular form of the more general law we call **Quanto-Geometry**. In previous discussion, we have stipulated and hereby reiterate that all objects are made up of mass and space. We have stipulated as well that they both influence each other to give birth to Quanto-Geometry. That is to say, in the process of mutual interaction, mass becomes intrinsically distributive and space becomes extrinsically geometric. We have equally stated that there exists no pure mass and no pure space in the universe because all apparent discrete masses and all apparent pure space are always coupled to one opposite gender partner. Quanto-Geometry is as such generalized in the universe and represents the fundamental modality of all forms of existence in the universe.

In giving a formal Quanto-Geometric account of an object beyond the generic momentum, we may elect to express independently what happens to "pure" mass due to its coupling to space and likewise what happens to "pure" space due to its association with mass. (The term pure here is relative, it means the pre-birth state of the coupled mass and coupled space as well). The exercise of the articulation is such that mass then becomes *spatialized* or distributive and space becomes *compressed* or *tensorized*. If we establish q as the quantum (scalar) value and s as the compressed space value, per previous notation, the Quantum part of the object is:

$$Q_2 = q \times s \quad (2\text{-}5)$$

We call this quantum part Q_2 because it is the product of an initial state Q_1 where the mass and the space deployment were "purer". We may then conceive that the quantum part Q_1 of the object involves an integrative sum, and the space deployment as well. We have earlier posed that space, albeit homogeneous, is variable. Therefore the variable of integration will be space. The quantum quantity will be a variable of space, and the space quantity will be a variable of *itself*. So that the initial or "purer" state of the quantum will be:

$$Q_1 = q \int s \cdot ds$$

$$= q \cdot \frac{s^2}{2} + C_q$$

It is important to comprehend the physical meaning of the integration of s with respect to itself. Space, being nothingness, is a continuum. The continuum does not disappear while space is in dilation or in propagation. This basically means that space can always be referenced to itself. That implies s to be a function in or of itself such as $s(s^*)$, where s^* is space as the independent variable and s the dependent variable or the function. In this development however, we are going to keep s as the independent variable of the $Q(s)$ function.

As to the quantum part q of the object, some might be uneasy with the representation of a macroscopic object holding a distributive mass, so deserving to be treated as a quantum. We do know however that the mass of any macroscopic object engaged in a non-linear motion, for instance, loses its point-like character and acquires inertia. The mass must then be related to the rotational characteristic of the object and become as such a *moment of inertia I*. We so take into account how the mass is distributed around an axis of symmetry of the motion and no longer speak of mass as such but of the quantity of inertia of the body.

Let us now take a look at the situation of the space spread part of the object in the coupling process. Since space experiences a compression within the articulation due to the constraint exerted by the mass partner, a formal expression for this compressed state must include a quantum factor or a coefficient of compression we call k. So that:

$$S_2 = k \times s$$

and the primary state of the space deployment of the object comes to be:

$$S_1 = k \int s \cdot ds$$

$$= k \cdot \frac{s^2}{2} + C_s$$

The expressions for Q_1 and S_1 are in fact operators. In analyzing the constitution of an object, these expressions formulate the request to square the variable part, divide this squared quantity by 2 and then multiply it by the involved constant, which may even be a coefficient. This operation is very common in physics and it becomes possible and meaningful only because of the generalization of Quanto-Geometry. This form of expression directly applies to the Quanto-Geometric composition of all universal objects. Let us now revisit the utilization of this form of expression throughout the entire discipline of physics.

a.- The linear and rotational Kinetic Energy

$$K_1 = m \cdot \frac{v^2}{2}, \quad K_r = I \cdot \frac{\omega^2}{2}$$

In these two equations, the variables are v, the linear velocity in a linear motion, and ω the angular velocity in non-linear motion. The coefficient m stands for the mass or, if you will, the translational inertia of the body in motion and I for its rotational inertia. They play both the role of the quantum part of the object. K_l and K_r are the respective kinetic linear and rotational energy of the object in motion.

b.- *The energy stored in the magnetic field of an Inductor*

$$U = L \cdot \frac{i^2}{2}$$

The variable here is the current i, which is squared, divided by 2 and multiplied by the inductance L to produce the energy stored in the magnetic field of the inductor.

c.- *The energy density in an Inductor*

$$U = \mu_0 n^2 \cdot \frac{i^2}{2}$$

The variable is still i in this equation, the constant or quantum part is $\mu_0 n^2$. The quantity $\mu_0 n^2$ being a product of the permeability μ_0, a constant and the number of turns n of the inductor, equally a constant, remains a constant as a whole and plays throroughly the quantum role.

d.- *The energy stored in a Capacitor*

$$U_1 = C \cdot \frac{V^2}{2}$$

In this situation, the variable is the voltage V across the plate of the capacitor and the quantum part is played by C, the cumulative capacity of the Capacitor called the capacitance.

e.- *The energy density in the electric field of a charged Surface*

$$U = \varepsilon_0 \cdot \frac{E^2}{2}$$

E, the electric field between the plates of the capacitor, is actually the variable in this situation, and ε_0, the permittivity constant, exhibits the quantum part.

f.- The kinetic energy of a fluid in motion

$$K = \sigma \cdot \frac{v^2}{2}$$

σ stands for the fluid density constant and v the velocity of the fluid, a variable. This equation is used in fluid mechanics and is part of Bernoulli's condition.

g.- The elastic potential energy of a spring submitted to a force

$$U = k \cdot \frac{x^2}{2}$$

U is the potential energy at any point of the spring. It depends on the distance x of the point from its initial position. This distance is obviously a variable. The spring constant k exhibits the quantum part here.

There may be more of these relationships than the above reported ones. It is all the more surprising that this form of expression applies to the kinetic energy and the potential energy alike. From our usual kinematic point of view, kinetic energy and potential energy are two radically different physical realities. The similarity of their expressions strongly suggests an equivalence between the variables they hold as well as the constants. That is to say velocity, current, voltage, electric field and distance would be, if not equivalent, yet of the same origin. Since we know velocity and current as definite representatives of movement, the above implies that the variables of voltage, electric field and distance (or space) must hold movement alike. Similarly, inertia, capacitance, energy density, permittivity and inductance are implicated to be of the same nature, if not equal. The best way to account for all of these symmetries is through the Quanto-Geometric premises set forth in the study in course. By the same token, we are pressed to thoroughly review our common notion of energy as well.

First of all, we ought to understand that the static form of the energy is not the potential but instead what we have commonly designated the kinetic energy, the term that includes the scalar quantity. The kinetic energy is very much indeed the potential or static form of the energy. Secondly, considering space as an integral part of the object, the energy is not an intangible quantity or entity hosted by the physical objects but instead an ontological part of the object. Thirdly, the energy is not a temporal quantity such that at one moment the object may hold it in a potential form and one instant later release it as kinetic energy. The physical object is always a sum of potential and kinetic energy concomitantly. The state of energy is the primary level of existence of the object. This is formally expressed in Quanto-Geometry as:

$$\varphi = \langle S_1 + Q_1 \rangle = q\int s \cdot ds + k\int s \cdot ds$$

$$= (q + k)\int s \cdot ds$$

$$= (q + k)\frac{s^2}{2} \quad (2\text{-}6)$$

We now differentiate equation (2-6):

$$\frac{d\varphi}{ds} = (q + k) \cdot s \quad (2\text{-}7)$$

Equation (2-7), with q being a scalar and k a scalar coefficient, expresses a momentum Q that we rewrite in the form:

$$Q = (q + k) \cdot s \quad (2\text{-}8)$$

So that by replacing (2-8) in (2-7):

$$\varphi = (q + k) \cdot \frac{Q^2}{2(q + k)^2}$$

$$\varphi = \frac{Q^2}{2(q + k)}$$

The scalar sum of (q+k) may be rewritten as:

$$M = q + k \quad (2\text{-}9)$$

To finally yield by replacement:

$$\varphi = \frac{Q^2}{2M} \quad (2\text{-}10)$$

We have elected to do away with the constant of integrations C_q and C_s in equation *(2-6)*, because we are considering the object in native or ground state. In the excited state of the object, equation *(2-10)* for the expression of the object becomes:

$$\varphi = \frac{Q^2}{2M} + C_{sq} \quad (2\text{-}11)$$

The above development characterizes a set of definitions that frame the endogenous organization of an object. In some ways expression *(2-11)* stands for an approximation to the Hamiltonian operator. Equation *(2-11)* expresses a preconceived Quanto-Geometric abstraction that we may apply to objects in order to elucidate, not only their structure, but their behavior as well, because the quantum-space organization of an object reflects on its spatial behavior or kinematics.

Effectively the form of the term $Q^2/2M$ in the equation is a very common expression in the rendering of physical phenomena whereby Q corresponds to a momentum or other similar two-term product quantity and M a physical scalar. Reconsidering the above energy identities, we find that they may be transformed to yield the form of the rational term $Q^2/2M$. The following will make it even more apparent.

2.4.1 Linear Kinetic Energy

$K_l = m\ v^2/2,\ p = m \times v,\ K_l = p^2/2m,$

where p is the linear momentum of the object in motion.

2.4.2 Rotational Kinetic Energy

$K_r = I\ \omega^2/2,\quad H = I \times \omega,\ K_r\ =\ H^2/2I,$

where H is namely the rotational momentum of the object in motion.

2.4.1 Energy in Magnetic Flux of Inductor

$U = L\ i^2/2,\quad Li = N\Phi,\ U = (N\Phi)^2/2L,$

where the quantity $N\Phi$ is defined as the flux linkage of the inductor.

2.4.2 Energy Density in an Inductor

$$U = \mu_0 n^2 \, i^2 \, /2, \quad \beta = \mu_0 n \, i, \quad U = \beta^2 / 2\mu_0$$

where β is the strength of the magnetic field set up by a current-carrying inductor.

2.4.3 Energy Stored in a Capacitor

$$U = C \, V^2 \, /2, \quad Q = CV, \quad U = Q^2 \, /2C,$$ Q is the total amount of charge stored on the plates of a capacitor.

2.4.4 Energy Density in Field of a Charged Surface

$$U = \varepsilon_0 \, E^2 \, /2, \quad \sigma = \varepsilon_0 \, E, \quad U = \sigma^2 / 2\varepsilon_0,$$

where σ is defined as the charge density on a surface.

2.4.5 Kinetic Energy in Liquid Flow

$$K = \sigma v^2 /2, \quad F = \sigma v, \quad K = F^2 /2\sigma,$$

where F is the flow rate of a liquid.

2.4.6 Elastic Energy Stored in a Spring

$$U = k \, x^2 \, /2, \quad F = k \, x, \quad U = F^2 /2k$$

The second equation is known as Hooke's law, and F the restoring force of a spring.

The difference between the incipient Quanto-Geometric operator above and the Hamiltonian lies in that the latter instructs about how to look at an oscillatory event in order to acquire knowledge about the characteristics of its variable, while the former screens objects (or events), whether oscillatory or not, from a scalar-space prism and directly expresses the structural order of the incipient observed object, which stands as a preview to its kinematics. In light of the unmitigated scope of the Quanto-Geometric bi-focal formalism, one shall give primacy to the Schrödinger's Hamiltonian of older Quantum Mechanics over Dirac's Lagrangian in the treatment of dynamics within the world of atoms. The latter shall be considered not as a substitute but an extension to the former, despite the two key merits of the Lagrangian. For purpose of reference, the extremely complex Dirac Lagrangian equation is reported in Appendix II.

2.5. *Quanto-Geometric Normalization*

Catering to the property of absolute homogeneity of space, the first law of Newton stating conservation of linear momentum *(p = mv)* has been viewed by and large as the most fundamental physical law. One might confer to the incipient Guanto-Geometric operator, which is based on the energetic distribution in an object, the standing of most fundamental law of conservation from the Quanto-Geometric perspective. Notwithstanding its primacy, it only represents one among many other forms of Quanto-Geometric articulation.

We will now study the functional expression dictating the specific forms of Quanto-Geometric articulation. At the outset, the expression of this function is as follows:

$$Q(s) = \pm \left| \frac{s^0}{\sigma\sqrt{2\pi}} \bullet e^{\left(\frac{-s^2}{2\sigma^2}\right)} \right|$$

Despite its relative similarity with the Gaussian, this function bears significant specifics solely proper to Quanto-Geometry. Its dependent variable is the quantum and is denoted $Q(s)$. Its independent variable is space and is denoted s. The $Q(s)$ function takes positive and negative values (Fig. 2.1). It is defined for all values of s except $s = 0$. When $s = 0$, a form of indetermination (0^0) appears in the numerator leaving the function undefined. The s-axis represents an asymptote. When s takes infinitely large values, $Q(s)$ approaches 0 without ever reaching it, converting the s-axis into an asymptote. This behavior of the function as conceived ensures that the Quanto-Geometric premises are fully respected. A primary postulate of Quanto-Geometry stated that no absolute mass and no absolute space can be present in the universe. We stipulated that pure mass was absolutely concentrated and pure space infinite or in infinite propagation. These unreachable absolutes, of which we can think nonetheless, can thus only stand for a reference, they may not be part of reality itself. The function perfectly models these conditions as well as the referential[2] or coordinate system does. We will later show further implications of the nature of the referential on the plane of number theory.

---

[2] In this exposé we use the terms referential and coordinate system interchangeably. We introduce the term referential to signify that we assign generic physical properties to the axes of the coordinate system beyond simple mathematical ordinalization.

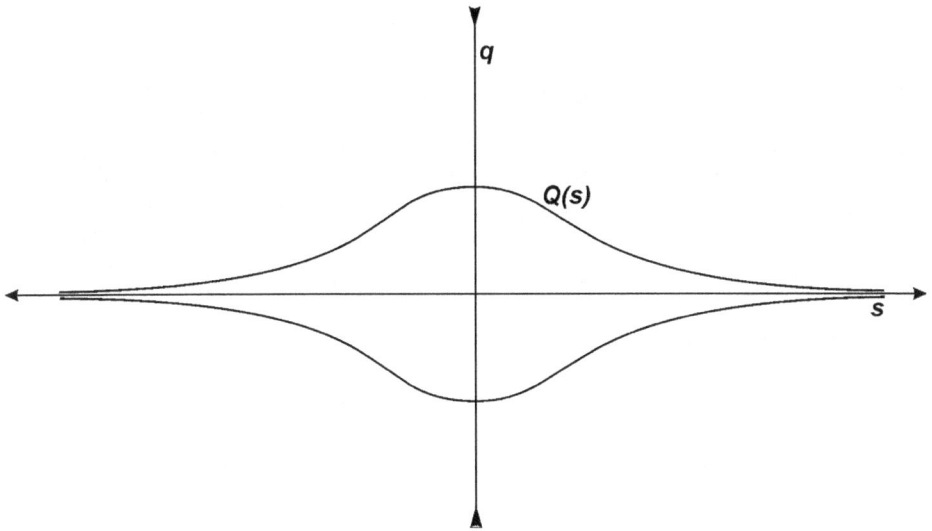

Fig. 2.1 Graph of the Quanto-Geometric function of one variable

The parameter σ stands for the accustomed standard deviation. It defines the evasiveness of the curve. Its real physical meaning is a coefficient of Quanto-Geometric articulation. Since $Q(s)$ has a discontinuity, thus formally speaking it is not a function and cannot be studied as such. To acquire the properties of a function, it must be expressed as shown below:

$$Q(s) = \pm \left| \frac{s^0}{\sigma\sqrt{2\pi}} \bullet e^{\left(\frac{-s^2}{2\sigma^2}\right)} \right|, s \neq 0$$

Notwithstanding the first part of the expression which looks even more similar to the Gaussian, the function is not the Gaussian. Most notably, because of the discontinuity at $s = 0$, this function calls for a different form of normalization than the conventional Gaussian. Its eventual mean value would fall within the area of discontinuity. For that matter, the modality of interpretation by the mean must be discarded. Consequently the variance also becomes meaningless. The function must be graphed on an orthogonal referential or coordinate system, with both axes equally normed, so that its inner geometric properties remain preserved. By no means can it be reduced to the ordinary "bell-shaped" curve. In a word, it is not a tool for the study of probability as we know it. Instead, it introduces a new form of normal distribution which goes beyond the Gaussian distribution, the Quanto-Geometric distribution.

The discontinuity constrains us to study the function either on the range]0, ∞] or]0, -∞] of the variable. We know that $Q(s)$ can never be equal to 1, the maximum attainable if there were no discontinuity. Starting in the neighborhood of 1, the function approaches 0 as s tends to infinity. In analyzing graphically the process of differentiation of this function, we notice that the tangential line to the curve, whose slope is nothing else than the rate of change dq/ds of the function, is external to the curve from the summit to the standard deviation point. As the function progresses downward to 0, the tangential line becomes internal to the curve. The rate of change dq/ds thus divides the function into two distinct sectors. The slope of the tangential line implies a relationship between s and $Q(s)$ that we will shortly investigate.

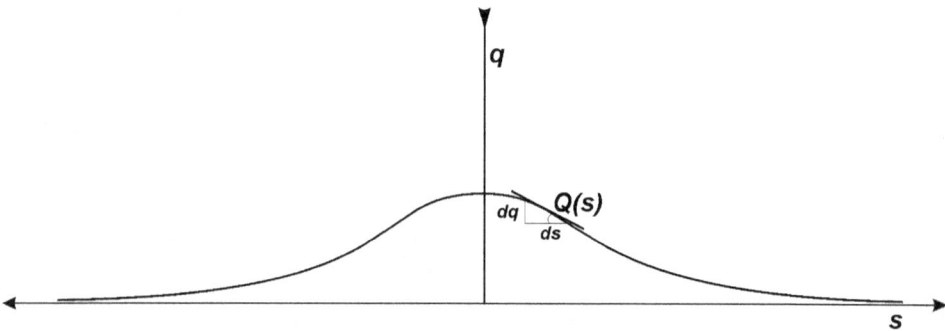

Fig. 2.2 Derivation by tangent line method

By constructing two lines, one parallel to the positive s-axis and the other parallel to the positive q-axis respectively, we form together with the tangential line a right triangle (Fig. 2.2). The slope of the tangential line is equal to the tangent of the angle θ.

$$\tan \theta = \frac{opposite\_side}{adjacent\_side} = \frac{dq}{ds}$$

We also note that the rate of change dq/ds is always larger than 1 from the summit of the curve to the standard deviation point:

$$\frac{dq}{ds} > 1$$

In this region the quantum differential is larger than the space differential. In other words, for a small change in space there is a larger change in the quantum. This means that the quantum imprisons space, it compresses it. Stated differently, within the articulation, the quantum is the dominant tenet. Sliding the tangential line down, we reach the *standard deviation* (inflection) point or σ where:

$$\tan \theta = \frac{dq}{ds} = 1$$

or $ds = dq$

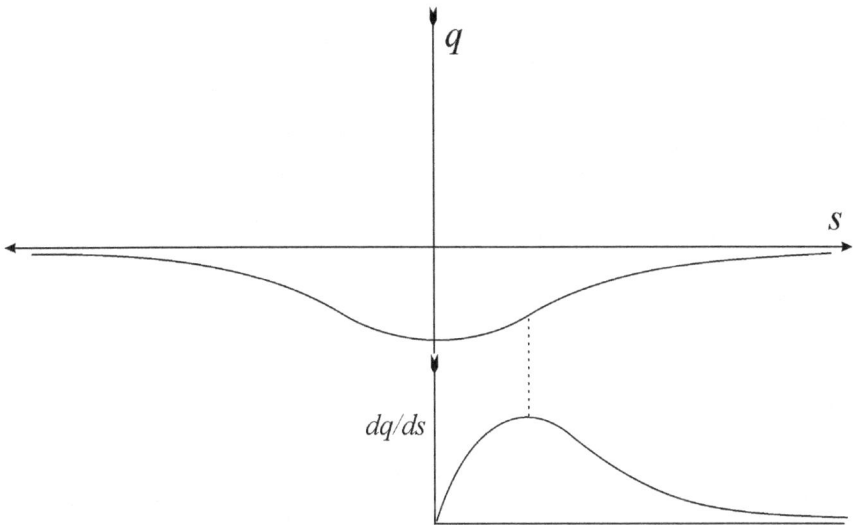

Fig. 2.3 Derivative of the Quanto-Geometric function

Here the differential *dq* is equal to *ds*, signifying that a change in space purports an equal change in the quantum. Within the Quanto-Geometric articulation at the standard deviation point, the constraint exerted by space on the quantum is equivalent to the one exerted by the quantum on space: mutual influence is equated at that point. In other words there is no resulting *imposition* on the part of any of the two factors. This is gathered in the null value of the second-order derivative at that point (Fig. 2.3):

$$\frac{d^2 Q(s)}{ds^2} = 0 \rightarrow s = \sigma$$

From the standard deviation point to infinity, the differential relation is:

$$\tan \theta = \frac{dq}{ds} < 1$$

$$\text{or } dq < ds$$

We can clearly read on the graph that the differential ds is larger than dq in this region. It is space now that jugulates the quantum. This physically means that the quantum becomes unapparent and submits itself totally to the degree of mobility of its space partner.

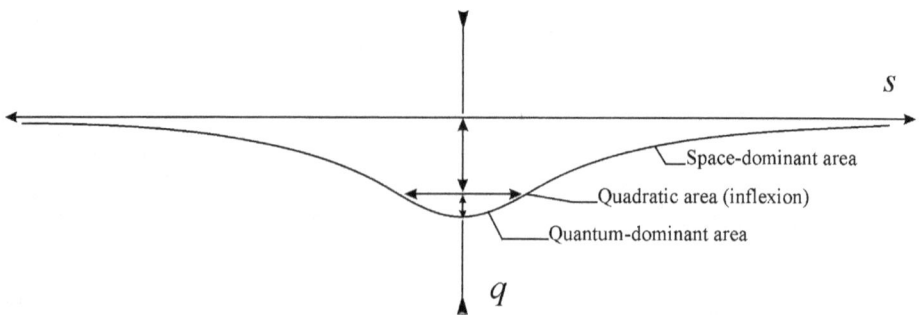

Fig. 2.4 Dissection of the Quanto-Geometric function in 3 main areas

We understand from the above analysis that, within the framework of our $Q(s)$ function, there exist three *quintessential* states of Quanto-Geometric articulation (Fig. 2.4). Further scrutiny of the curvature of the function tells us however that these states include 9 different granular forms of articulation. Each responds to a particular form of curvature contained within the geometry of the function. The following sequence, a paramount entity, will help us delineate these fractional forms of curvature:

$$\{q_n\} = \left\{ \frac{e^{(1-n)/(10-n)}}{\sigma\sqrt{2\pi}} \right\}, \quad \text{where } 1 \le n \le 9 \quad (2\text{-}12)$$

This sequence is of capital importance, surpassed only in significance by the following pilot geometric sequence which it stems from:

$$\{s_n\} = \left\{ \frac{1-n}{10-n} \right\}, \quad \text{where } 1 \le n \le 9 \quad (2\text{-}13)$$

Those sequences determine the higher norm of **symmetry-breaking** that ultimately presides over the development of all objects modeled by the Quanto-Geometric Function. The $\{q_n\}$ sequence sets the end point of each specific form of symmetry embedded in the function throughout its graphical course, so defining the intervals of Quanto-Geometric *norm-alization*. This sequence defines the set of real numbers which will dissect the curve of the $Q(s)$ function into its 9 *natural* discrete forms of curvature, whereas the $\{s_n\}$ sequence determines the pivotal s points on the space line that give rise to the specific s domains of normalization. So therefore, the first sequence yields the values taken by the function at the points of separation of curvature sectionals, the defining points of symmetry-breaking. By replacing n with the values in its defined range, the resulting terms of the sequence follow (see section 4.6 for relationship between n and s):

$$\text{for } n = 1, \quad \frac{e^0}{\sigma\sqrt{2\pi}} = \frac{1}{\sigma\sqrt{2\pi}} = Q_1(s)$$

$$\text{for } n = 2, \quad \frac{e^{-1/8}}{\sigma\sqrt{2\pi}}, Q_2(s)$$

$$\text{for } n = 3, \quad \frac{e^{-2/7}}{\sigma\sqrt{2\pi}}, Q_3(s)$$

$$\text{for } n = 4, \quad \frac{e^{-1/2}}{\sigma\sqrt{2\pi}}, Q_4(s)$$

$$\text{for } n = 5, \quad \frac{e^{-4/5}}{\sigma\sqrt{2\pi}}, Q_5(s)$$

$$\text{for } n = 6, \quad \frac{e^{-5/4}}{\sigma\sqrt{2\pi}}, Q_6(s)$$

$$\text{for } n = 7, \quad \frac{e^{-2}}{\sigma\sqrt{2\pi}}, Q_7(s)$$

for $n = 8$, $\dfrac{e^{-7/2}}{\sigma\sqrt{2\pi}}, Q_8(s)$

for $n = 9$, $\dfrac{e^{-8}}{\sigma\sqrt{2\pi}}, Q_9(s)$

By substituting these values of the function in its equation, we compute the corresponding values of the variable s:

$n = 1$, $s_1 = 0$

$n = 2$, $s_2 = \pm\,\sigma/2$

$n = 3$, $s_3 = \pm\,2\sigma/7^{1/2}$

$n = 4$, $s_4 = \pm\,\sigma$

$n = 5$, $s_5 = \pm\,\sigma(8/5)^{1/2}$

$n = 6$, $s_6 = \pm\,\sigma(5/2)^{1/2}$

$n = 7$, $s_7 = \pm\,2\sigma$

$n = 8$, $s_8 = \pm\,\sigma7^{1/2}$

$n = 9$, $s_9 = \pm\,4\sigma$

The value of the sequence for $n = 1$ determines a value of the Quanto-Geometric function which is in fact a limit:

$$\lim_{s \to 0} Q(s) = 1$$

At $s = 0$, the corresponding value of the variable for $n = 1$, the Quanto-Geometric function approaches a limit. It is important to realize that the function is not defined at $s = 0$, even though our graphism of representation thus far does not portray that discontinuity, simply for practical graphing purposes. In the interval $]0, \sigma/2[$, the curvature of the graph is circular or **monocentric**. In the next interval $[\sigma/2, 2\sigma/7^{1/2}[$, the geometry of the curvature is elliptical or **bipolar**. The section of curvature laying in the interval $[2\sigma/7^{1/2}, \sigma[$ is parabolic or **associative**. In the interval $[\sigma, \sigma(8/5)^{1/2}[$ the function is of hyperbolic or **ambivalent** curvature. The forms of curvatures we know as conic sections end up with this interval. In the fifth interval $[\sigma(8/5)^{1/2}, \sigma5^{1/2}[$ the function displays a form of curvature we call the **entropic** form of curvature. The sixth interval $[\sigma5^{1/2}, 2\sigma[$ delineates a form of curvature we denote the **enthalpic** curvature. In the seventh interval $[2\sigma, \sigma7^{1/2}[$ the function describes a form of curvature that we call **paradoxical** curvature. The eighth interval $[\sigma7^{1/2}, 4\sigma[$ encompasses a fraction of curvature we coin the **perfect** curvature. And finally in the ninth interval $[4\sigma, \infty[$ the function displays a section of curvature we denote linear or **infinite** curvature. The choice of these attributes will be completely justified later in this exposition.

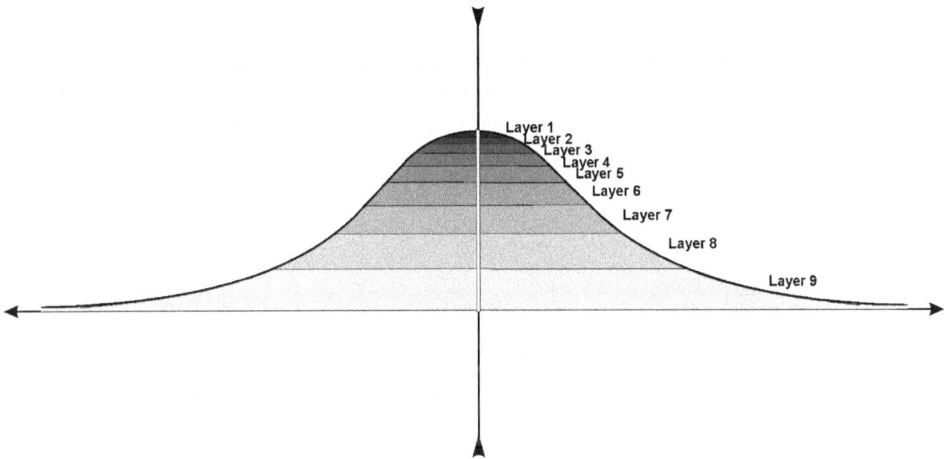

Fig. 2.5 The 9 essential layers of the Quanto-Geometric function

These intervals define the *normalization* modes of the Quanto-Geometric function. Taken horizontally, they define the different levels of Quanto-Geometric articulation. The areas that make sense under the curve are not vertically oriented as the normal Gaussian distribution has it. The meaningful and useful areas, from the Quanto-Geometric perspective, are the ones horizontally delineated by the natural quantum values of the function (Fig. 2.5). These areas, termed layers in the above illustration, are Quanto-Geometric *normals*. One way to visualize these levels, from a conventional modern physics standpoint, is in terms of energy levels. We have seen earlier that what we call energy is the structural expression of an object. We may now understand that our usual concept of mechanical energy (kinetic + potential) corresponds in fact to the first level of the Quanto-Geometric function or the first *normal*. In Quantum Mechanics, as the energy level increases the particle is able to remain preferably farther in space than its regular "ground" level location. Our model readily explains this effect with the visualization that the higher the Quanto-Geometric level displayed in the graph of the function, the larger the related space extent s or the *normal*. We also know, in Quantum Mechanics, that the energy associated with a level is always given as a difference between two quantities. This scheme is part of the Quanto-Geometric model on account of the area included in one horizontal stripe. In comparison with the energy levels, the size of that area is equal to the difference between a larger area bounded by the upper limit of the interval and a smaller area bounded by the lower limit of the interval.

On another hand, our model readily explains why there is a specific geometry associated with each energy level. We will not, however, formally utilize the term energy in our exposition, since each level represents indeed a preeminent physics law in the light of Quanto-Geometry.

2.6. The Nine Quanto-Geometric Primitives

If it indeed holds true that the Quanto-Geometric function contains the conservation operator, in the form that we have expressed it in section 2.4, we must be able to retrieve the operator from the expression of the function. Let us start by writing the function in an implicit form by replacing the notation Q(s) by q, in the same way that we traditionally replace the notation f(x) by y. The Gaussian-like part of the Quanto-Geometric function may then be written as:

$$q = \frac{1}{\sigma\sqrt{2\pi}} e^{\frac{-s^2}{2\sigma^2}}$$

$$q\sigma\sqrt{2\pi} = e^{\frac{-s^2}{2\sigma^2}}$$

$$\frac{-s^2}{2\sigma^2} = \ln\left[q\sigma\sqrt{2\pi}\right]$$

$$\frac{-s^2}{2\sigma^2} = \ln q + \ln \sigma + \ln\sqrt{2\pi}$$

$$\frac{-s^2}{2\sigma^2} = \ln q + \ln \sigma + \frac{1}{2}\ln(2\pi)$$

$$\frac{-s^2}{2\sigma^2} = \frac{1}{2}\left[2\ln q + 2\ln \sigma + \ln(2\pi)\right]$$

$$\frac{-s^2}{\sigma^2} = 2\ln q + 2\ln \sigma + \ln(2\pi)$$

By multiplying numerator and denominator of the first term by $2q^2$:

$$\frac{-s^2}{\sigma^2}\cdot\frac{2q^2}{2q^2} = 2\ln q + 2\ln \sigma + \ln(2\pi)$$

$$\frac{-s^2 q^2}{2q}\cdot\frac{2}{q\sigma^2} = 2\ln q + 2\ln \sigma + \ln(2\pi)$$

$$\frac{-s^2 q^2}{2q}\cdot\frac{2}{q\sigma^2\left[2\ln q + 2\ln \sigma + \ln(2\pi)\right]} = 1$$

$$\frac{-(sq)^2}{2q}\cdot\frac{2}{q\sigma^2\left[2\ln q + 2\ln \sigma + \ln(2\pi)\right]} = 1$$

Replacing sq by Q ($Q = s \times q$ from equation 2-1):

$$\frac{-Q^2}{2q}\cdot\frac{2}{q\sigma^2\left[2\ln q + 2\ln \sigma + \ln(2\pi)\right]} = 1$$

Let: $2\ln q = a$,

then $\dfrac{a}{2} = \ln q$

and $q = e^{\frac{a}{2}}$.

Let: $2\ln\sigma = b$,

then $\dfrac{b}{2} = \ln\sigma$

and $\sigma = e^{\frac{b}{2}}$.

Let: $\ln(2\pi) = c$

The equation then becomes:

$$\frac{-Q^2}{2q} \bullet \frac{2}{e^{\left(\frac{a}{2}\right)} \cdot e^{\left(\frac{b^2}{4}\right)}\left[a+b+c\right]} = 1$$

$$\frac{-Q^2}{2q} \bullet \frac{2}{e^{a/2} \cdot e^{b^2/4}\left[a+b+c\right]} = 1$$

$$\frac{-Q^2}{2q} \bullet \frac{2}{e^{\left(\frac{2a+b^2}{4}\right)}\left[a+b+c\right]} = 1$$

$$\frac{-Q^2}{2q} \bullet \frac{2}{ae^{\left(\frac{2a+b^2}{4}\right)} + be^{\left(\frac{2a+b^2}{4}\right)} + ce^{\left(\frac{2a+b^2}{4}\right)}} = 1$$

First, notice that the equation holds true only if $s \neq 0$, which is an explicit condition of the Quanto-Geometric function. If s equals 0, then Q equals 0 and we end up with the absurd result of $0 = 1$. So therefore s cannot be equal to 0. In its usual form, the Gaussian function does not incorporate this condition, for it is defined for all values of s. It is all the more significant that when the Gaussian expression, as a part of the Quanto-Geometric function, is rewritten in an implicit form the condition pops out. Even though unapparent, the condition is therefore implicit in the Gaussian function as we convention-ally know it, and further poses significant implications regarding the physical meaning of the variable s. These implications are but the Quanto-Geometric premises set forth at the beginning of this discussion.

Also notice that the equation is equal to 1, a condition commonly known as the condition of normalization. It means that the equation expresses the totality of reality, or the entirety of the particular physical phenomenon which it is modeling. Totality means 100%, that is the ratio 100/100, which is equal to 1. We can perfectly replace 1 by ψ because we Quanto-Geometrically think of this totality as an object, in the same manner as we did for the former Quanto-Geometric operator equation. We then express the above equation in the following fashion:

$$\frac{-Q^2}{2q} \bullet \frac{2}{ae^{\left(\frac{2a+b^2}{4}\right)} + be^{\left(\frac{2a+b^2}{4}\right)} + ce^{\left(\frac{2a+b^2}{4}\right)}} = \left\| \Psi \right\| \quad (2\text{-}8)$$

The characteristic term $Q^2/2q$ is again present as is and needs no further consideration. The term $2/(\,ae^{(2a+b^2)/4} + be^{(2a+b^2)/4} + ce^{(2a+b^2)/4})$, expresses the specificities of a relationship between the quantum q of circulation and the standard deviation value σ of the space under tension. Exclusively made up of q and σ, the exponential term represents in fact the footprint of a transcendental function of the hyperbolic type. As we know, the hyperbolic functions are generally based on a ratio between the number 2 and a sum of exponentials, commonly two of them and each raised to an x and an $-x$ power (x being the independent variable). The transcendental part of the above (2-8) expression is a general formulation for the precursor of the family of hyperbolic functions and contains all of the possible forms of symmetry. In support of this claim, one should remain wholly mindful that there is knowingly complete correspondence between the family of trigonometric functions and a subset of the family of hyperbolic functions.

In addition, there is a set of values for parameters a, b and c in the (2-8) equation that will give us 9 different normalized values of $\left\| \Psi \right\|$, such as $\left\| \Psi_{g1} \right\|$ ... to $\left\| \Psi_{g9} \right\|$ all representing the 9 prime Quanto-Geometric Operators. However we will not pursue operator-based analysis in this text, remaining in favor of functional analysis in a bid to avoid the *uncertainties* generally associated with Operator-based analysis. We must keep in mind that an Operator is essentially a pre-conceived notion about a system, the solutions cast by that treatment being as good as the abstractions implicated by the Operator. We will make use of Operator treatment only when absolutely necessary. Furthermore, in that event we will use explanatory language instead of the nomenclature reserved to Operator-based analysis in an effort to keep this text accessible to the largest audience possible. Alternatively, we will equally refer to the 9 Quanto-Geometric normals in terms of **Primitives**.

In sum, we have established so far a certain number of Quanto-Geometric principles, proposed a Gaussian-like functional expression of these principles and demonstrated the close equivalence between the implications of the Gaussian and the Quanto-Geometric platform.

2.7. Quanto-Geometric Tri-Valent Operator

The examination of the geometric progression of the curve of the Quanto-Geometric function has revealed 3 grand moments or phases of the function. One where the value of the derivative was larger than 1 *(dq/d s> 1)*, another where the value of the derivative was equal to 1 *(dq/ds = 1)* and lastly one where the derivative was less than 1 *(dq/ds <1)*. This is a result of the *quintessential* Quanto-Geometric articulation incarnated in the function. This tri-partite phasing of the Function goes beyond functional mathematical representation and is best expressed in terms of mathematical operators. We gather this concept therefore under the term Quanto-Geometric Tri-Valent Operator, in the understanding that it is an Operator that encompasses 3 sub-operators. The first of these sub-operators we designate the Quintessential Quantum Operator or Hyper Quantum Operator (Fig. 2.6). The second of the sub-operators we coin the Quintessential Quadratic Operator or Hyper Quadratic Operator. The third of these sub-operators we dub the Quintessential Space or Hyper Space Operator. Whenever they participate in the configuration of matter, these operators do not each act alone or individually, they all three intervene concomitantly. We have given them the full designation of Operators, because the transformations they operate are first-order Operator's transforms. However it must remain completely clear that they each belong to the larger entity we designate the Tri-Valent Operator.

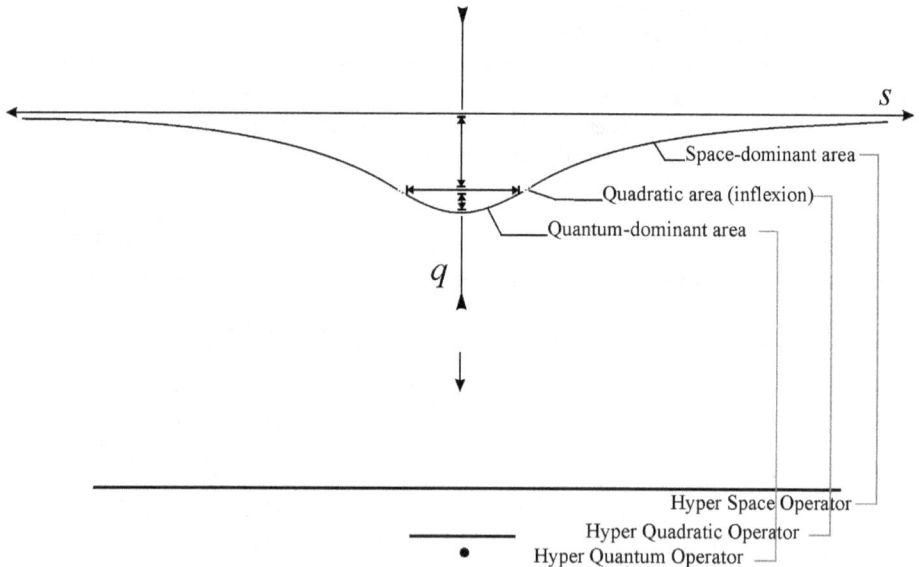

Fig. 2.6 Quanto-Geometric Quintessential Trivalent Operator

The *norm* of the Hyper Quantum Operator is the perpetration of utmost *quantification* or *densification,* and the overloading of everything pertaining to the signature of Quanto-Geometric type 1. The *norm* of the Hyper Quadratic Operator is the perpetration of utmost *ambi-valence* and the overloading of everything pertaining to the signature of Quanto-Geometric type 4. And finally the *norm* of the Hyper Space Operator is the perpetration of utmost *spatiation* and *spatialization* and the overloading of everything pertaining to the signature of Quanto-Geometric type 9. The details of these signatures we shall study later in the exposé addressing Quanto-Geometric Number Theory. As a reminder, the *norm* of a mathematical operator is the typical feature or behavior that the Operator propitiates.

2.8. An Alternative Notional View of Probability Theory

Before we proceed with the analysis of the other Quanto-Geometric primitives, let us briefly revisit the statistical notion of probability. We now know for certain that the Gaussian function is much more than what it purports to be at face value. Through Quanto-Geometry we have raised it to the category of a model capable of describing not only probability distribution but the physical universe and all of its contents. The paramount reason why we have created the concept of probability in parlance and mathematics is our ignorance of the essential nature of space and its phenomenological behavior. We look at the contents of the universe as discrete components and whenever we find ourselves unable to identify a causal relationship accountable for an observable physical effect, we interpret the latter as a statistical phenomenon permeable to intelligence only through the law of probability.

Let us now consider an ordinary statistical phenomenon. Gather any number of individuals in a room, give them a ruler and assign them the task of measuring the length of any of the walls of the room. You will find that the number of individuals versus of the measured length values describe a Gaussian distribution. In other words, most of them will hold values that are close to a specific one, namely the mean. As you consider a lower and lower number of individuals with same measurements, you will find that their error, or the dispersion around the mean, becomes larger and larger. That is to say, few individuals will have extremely large margins of error. In short, the correlation between a number of these individuals and the length they measured for the wall is modeled by the Gaussian function. Take a value of measured length, make it equal to the variable and you will find a value of the function yielding directly the number of individuals who measured that same length. Notwithstanding the beautiful mathematical operation allowed by the function, it does not explain why, no matter the number of individuals, no matter their temper, their race or social characteristics, no matter the size of the room, independently of everything else, the correlation between the measured values and the number of individuals with same measured values always holds the same form, or the same form of dispersion.

If we look at the phenomenon "individuals measuring the length of a wall" in the light of Quanto-Geometry, we may obtain a better understanding of the phenomenon than a mere statistical account. Quanto-Geometry poses that these individuals are not really discrete, absolutely separated from each other. They are made up of quanta, which unfortunately are the only parts we can perceive of them, and of space, which we unfortunately always disregard. They interact with each other even though they don't hit each other or they don't speak to each other. They intrinsically interact through their space component, and for that matter they can be together visualized as a whole or one individual. They make up a Quanto-Geometric object, that is to say a space field inhabited by the quanta, their observable bodily parts. Whatever they do "as a whole or in common", whatever they do "together", they will do it according to the fundamental Quanto-Geometric articulation for any object. There will be a point of quantum-space articulation where the value of the inertia is extremely high and jugulates space. These individuals are very much "localized" and in an event of measurement their deviation will be minimal producing the most accurate measurement. Notice that the prediction of the model is that the inertia will be the highest at that point. It certainly does not mean that the heaviest individual will occupy this level of the distribution but that there will be many individuals in this particular group. Remember that inertia means to us distributive mass akin to the quantum and we have made these two terms interchangeable. At the other extreme, we have a point of articulation where the space component jugulates the inertia. These individuals are very dispersive and, in an event of measurement, the deviation from the mean will be so significantly high that the measurement will be extremely inaccurate. The quantum-individuals making up this level of the "object", which is the highest one, will be scarce: the quantum is overwhelmed, it is unapparent, there is "almost" no quantum in this state. In the intermediate position, there is a point of articulation where the quanta are in equalization with the articulated space, these individuals are in the quadratic number and they measure the wall with the "standard" deviation σ. All others follow the Quanto-Geometric pattern alike in terms of their grouping and the dispersion in their measurements.

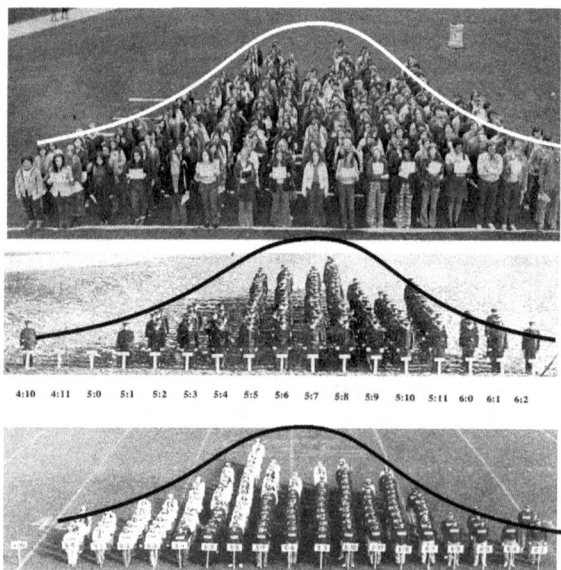

Fig. 2.7 Examples of height distributions in limited random samples of population

Quanto-Geometry further tells us that there will never be one group of individuals to measure the wall with 100% accuracy. We can fully take stock of the adequacy of this assumption because no measurement can admittedly ever be done with 100% accuracy. The real length of the wall is a limit that can only be approached. Quanto-Geometry also tells us that no matter how much deviated a measurement turns out to be, it is never wrong. There will always be at least one group of two individuals with an extremely deviated value, but they are never totally wrong. This is an irrevocable Quanto-Geometric stipulation. So it is because the Quanto-Geometric vision assesses that every object possesses a space component which is limitless, notwithstanding its level of compression or tension. As such the object is in essence not measurable or delimitable in the absolute.

This is indeed an example of the presence of quantization right in the midst of the macroscopic world, among a myriad of others such as the height distribution in human population (Fig. 2.7). Many may wonder how a theory describing the quantum world of particles may yield interpretative elements for the macrocosm, knowingly a continuum domain. By allowing us to divide the individuals of our example into specific groups of inertia, which undoubtedly correlate with the fundamental quantum numbers, this treatment has demonstrated how quantization works in the macrocosm in spite of its apparent continuum. Each of these individuals, no matter the kind of actions taken while performing the measurements, (one might have stopped for some coffee, while another might have stepped aside for a cigarette or some vapor), was part of a quantum group, or a group of inertia if you will, which belongs in a specific Quanto-Geometric level. While they were moving up and down busy at their measurements, you could have used Newtonian physics to measure their weight, the friction of their shoes against the floor, the pitch of the sound of their voices, etc., nevertheless they were all involved in a higher frame of event where they were in fact quantized. If you do not invoke space and its articulation with the quantum-bodies, there will be no possible explanation for such a peculiar and perhaps magical form of reality, which we call the Gaussian distribution, right within the macroscopic world, and we will be left at the end only with a statistical account of the event as is usually done. In the last Chapter of this text, we provide a more formal framework to this view.

The hidden variables invoked by A. Einstein to sustain an objection to the philosophy implied by Quantum Mechanics are no less than the Quanto-Geometric principles. The basic problem of Quantum Mechanics, a science heavily based on probability density distribution, dwell not in its mathematical results, which have always been universally accepted, but in the interpretation of these results. The uncertainty in the philosophical implications of the science of quanta had sparked a debate which divided physicists in two schools, the school of Copenhagen including among others Neils Bohr, Max Born, Paul Dirac, Wolfgang Pauli and Werner Heisenberg, and the school of determinism including Albert Einstein, Nathan Rosen, Boris Podolsky and John Wheeler among others. They have all strongly contributed to the onset of Quantum Mechanics however. While the school of Copenhagen developed a philosophy of uncertainty as a prevailing implication of Quantum Mechanics and as an intrinsic component of reality, the other stream fiercely resisted this view advocating the existence of unknown or hidden variables accountable for physical effects with yet no recognizable causes. They termed Quantum Mechanics an incomplete science insofar as an explanation of what makes physical objects or events behave or occur statistically remained to be found. A. Einstein was even quoted saying that he cannot "imagine the electron hopping like a bug", like the individuals of our previous example, that is!

We hope to have presented thus far enough novel elements of interpretation for the reader to at least perceive where the truth lies. The panorama will become much clearer after the study of the remaining eight Quanto-Geometric operators, and the kinetic-symmetric properties of space within each operator as well.

2.9. The Quanto-Geometric Coordinate System

The Quanto-Geometric coordinate system encompasses the fundamental concepts of space and mass at the most generic level that we may possibly visualize them (Fig. 2.8). Its vertical q-axis holds the regency of the scalar domain, the purview of the real number line, whereas the horizontal s-axis embodies the scale of void, both abstract and physical. The referential is orthonormed in relation to all axes. This referential or coordinate system is not a Cartesian coordinate system, although it may appear of great similarity with the latter. Further the s-axis may be represented either as a single line for a two-dimensional referential overall or as a double axis which together with the q-axis makes for a 3-dimensional referential. In its incarnation as a double axis, the s linear run makes explicit the two inner dimensions of void, which are symmetry and mobility as discussed later, thereby eliciting the Quanto-Geometric principle that void in its quality of a tenet of matter contains two degrees of freedom.

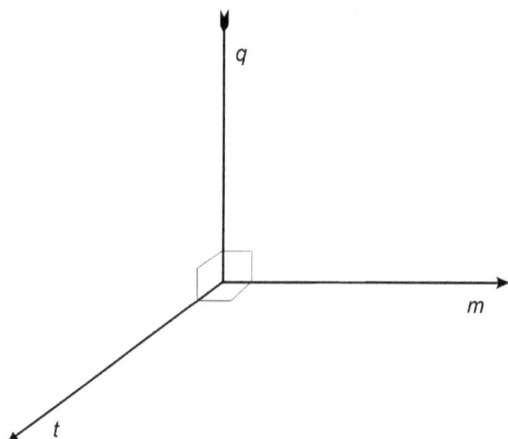

Fig. 2.8 Three-dimensional Quanto-Geometric coordinate system

What's more, while the q-axis is finite, the s-axis is infinite. That is the reason why the symbology of the q-axis portrays an inward-pointing arrow, whereas the s-axis portrays an outward-pointing arrow. The q-axis is engaged in the business of *discretisation* and the *infinitesimal*, in contrast to the regency of *continuum* and *infinity* endorsed by the s-axis (although restricted by symmetry-breaking intervention as we shall see later). The graphism we use in Quanto-Geometry may facultatively omit the depiction of the arrows as a matter of practicality. When it does however, one must keep in mind the nature of the principles embodied in the axes of the coordinate system. In its role of representative of the real number line the q-axis grandfathers the entire suite of real numbers from the largest integers to complex numbers to transcendental numbers. While one may easily grasp how integers, rational numbers as well as irrational numbers are modeled by a scaling entity such as a line, it may be less intuitive to visualize how that line may equally model or scale the set of complex numbers. The imaginary part of those numbers would presumably pose an obstacle to the scaling by the real axis of representation. One ought to remember the towering Quanto-Geometric principle that posits all scalar to be necessarily accompanied by a degree of void, no matter how minimal, as an existential partner. Consequently the most dense scalar cannot be devoid of a space component altogether. That is the reason why the real number line in Quanto-Geometry (in traditional mathematics as well I might add) is indeed a density referential. When it comes to the imaginary part of a complex number, it is no more than an expression of the void component of the number, which takes the embodiment of a geometric entity both in the rectangular form $(a + bj)$ of the complex number and in its exponential form $(ce^{j\theta})$.

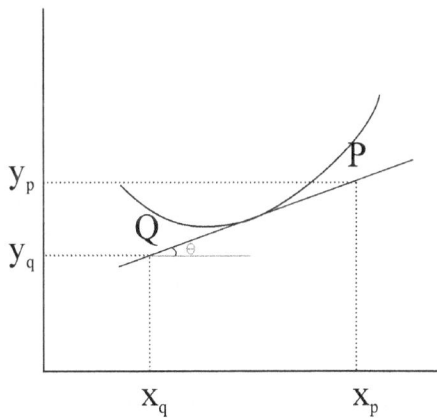

Fig. 2.9 Definition of a derivative at one point of a function in differential calculus

Effectively in differential calculus, incidentally also called *infinitesimal* calculus, the instantaneous rate of change of a curvature is equal to the tangent of the angle subtended by the curvature, expressed as below:

$$\frac{dy}{dx} = \tan\theta,$$

$$\text{so that:}\quad \theta = \tan^{-1}\left(\frac{dy}{dx}\right).$$

This θ angle is exactly the same as the one pertaining to the rectangular form of a complex number as well as the exponential form (Fig. 2.9):

$$\theta = \tan^{-1}\left(\frac{b}{a}\right)$$

$$\text{and } ce^{j\theta}.$$

As a reminder c relates to a and b by:

$$c = \sqrt{a^2 + b^2}$$

Consequently, it is important to keep in mind that the imaginary part of a complex number is equivalent to a form of the *infinitesimal*, and as such fully scalable on the traditional real number line and on the Quanto-Geometric q-axis. The above entails that the imaginary part of a complex number, if it must remain complex, cannot become 0 and understandably so. So does the affine increment ds in the ratio dq/ds, a condition that ultimately springs from the nature of the scalar itself. Although point 0 exists in the coordinate system on all axes, no function is allowed to take that value, both on the independent variable side and on the dependent variable side. It is a long standing debate in traditional mathematics as well whether or not 0 belongs to the real number line.

Undoubtedly, the Quanto-Geometric coordinate system is significantly different from the Cartesian coordinate system. We have made it a total referential that is capable of scaling all scalar quantities in a very explicit way. All axes must be orthonormed and normal to each other. Should that condition not be raised on the referential, the Quanto-Geometric functions would lose all of their quantum and geometric properties. They would become meaningless and unable to reflect graphically the inner quantum and geometric qualities innate to their respective expressions.

2.10. Quanto-Geometric Symmetry Groups

We are now going to introduce one last but not least significant descriptor to the Quanto-Geometric model. Because the physical variable of space purports dimensionality, any space descriptor must assume this essential property of *multiplicity*. One other way of looking at multiplicity is *variability*.

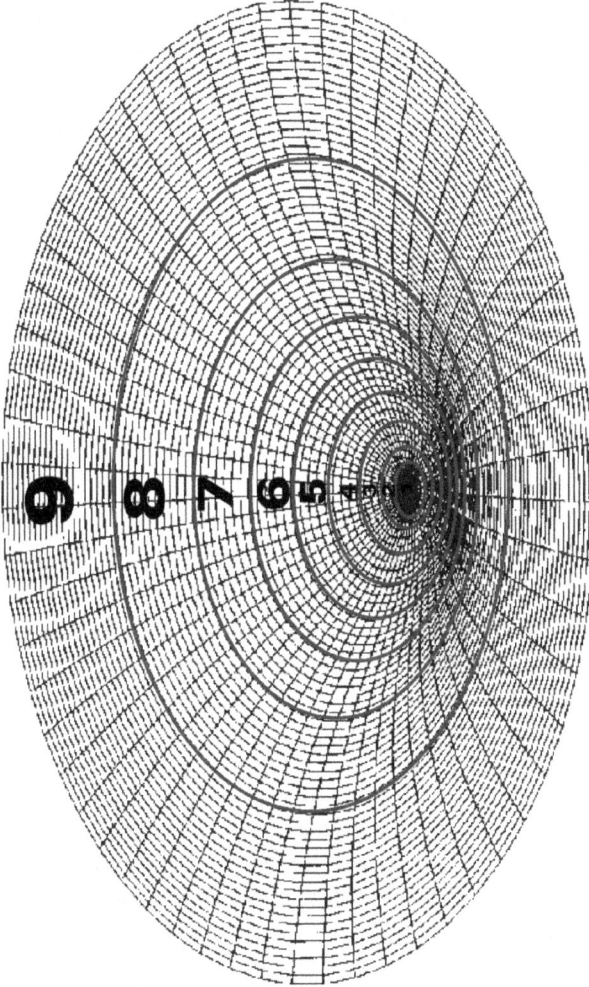

Fig. 2.10 The nine base shells of the Quanto-Geometric function of two variables.

For that matter, we have primarily set space as the independent variable of the function itself. However, we must further expand the expression of multiplicity embedded in the mathematical construct of *variable* by multiplying or surrogating the variable. Consequently, we must establish the Quanto-Geometric function as a function of two variables. The space variable s becomes as such a composite variable in order to incarnate the two basic sub-components of space: *form* and *mobility*. As we pointed out earlier, all forms are forms of symmetry. The end result, on the mathematical plane, is that the composite variable s may be cast into two separate variables, denoted m and t, the first standing for symmetry and the second for mobility. Our Quanto-Geometric function of one variable $Q(s)$ then becomes $Q(t,m)$, a function of two variables. We may appropriately rewrite the function as follows:

$$Q(m,t) = \pm \left| \frac{(m+t)^0}{\sigma \cdot \sqrt{2\pi}} \bullet e^{-(m+t)^2/2\sigma^2} \right|, \ (m,t) \neq (0,0), (m = \pm|t|) \, and \, (-m = \pm|t|) \ \text{(2-14)}$$

Although the same expression might be alternatively written in polar coordinates, we do not intend to perform any further development to the function in that direction. The variables t and m introduce two new dimensions to the Quanto-Geometric function, making it apt to be displayed in a three-dimensional referential with two independent variables substituting the one s variable. Albeit modeling each a different construct, these variables behave exactly alike. Shall we take a look at the graph of $Q(t,m)$?

We build the graph by holding constant one variable, t for example, assigning a value m_i to m, the other variable, and computing the value of $Q(t,m_i)$ (Fig. 2.10). When we do this for one constant value of t and successive values m_i of m, we obtain a Gaussian-like curve centered on that value of t, which is perpendicular to the t-axis and symmetric about it. By repeating the procedure with a span of t values, we obtain the three-dimensional graphical layout of the function.

The function is thus as symmetric about the t-axis as it is about the m-axis. This form of symmetry we generically call *transverse symmetry*. The elements in one quadrant will have common behavior among themselves, but each will have an opposite equivalent in the opposite quadrant, exactly like a 10° angle is transversally symmetric to a 170° angle. As one can see, these are descriptors derived from the trigonometry of the function. It could be the case that expressing the function in polar coordinates makes even more apparent the property of transverse symmetry. So instead of translating s as $s = (m,t)$ we might want to express it as $s = (r, \theta)$, where r is the magnitude of a desired radius and θ its angle with the s axis. So therefore, the point (r_1, θ) is transversally symmetric to the point $(r_1, \pi-\theta)$, about an s axis construed as identical to the trigonometric cosine axis. This is likely to be the technique used by the computer program that built the graph in Fig. 2.11. We see therein a series of concentric circles (different radii) around the q-axis and a series of top-down and bottom-up stripes (different angles) all across the circles.

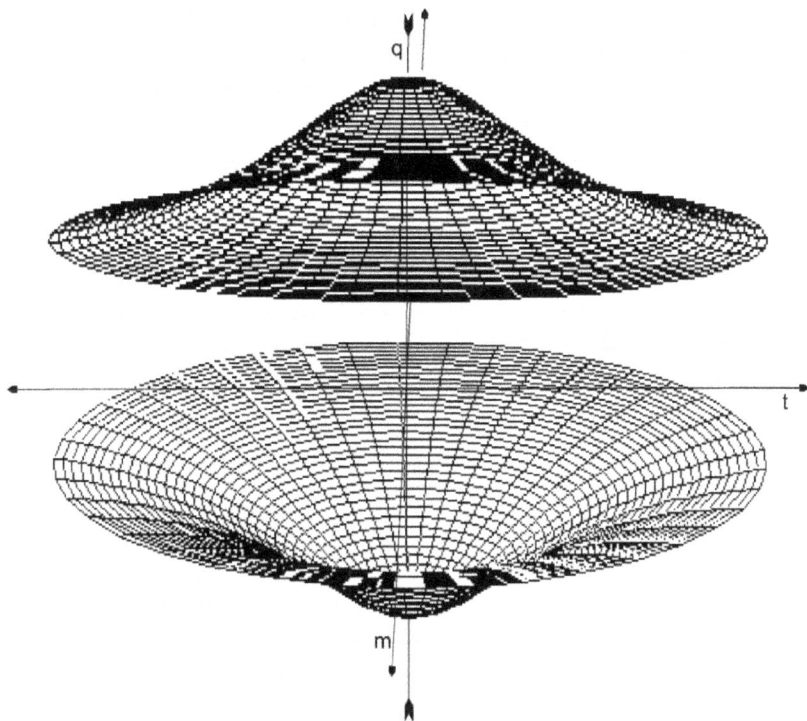

Fig. 2.11 A complete view of the graph of the Quanto-Geometric function of two variables.

If we look at the graph atop the vertical q-axis, we would see a series of 9 concentric stripes (washers) on an *m-t* plane (Fig. 2.12), every other stripe being larger than the previous as determined by the specific extent of the intervals of normalization. Hence, for every quarter of stripe the function exhibits a transversally symmetric quarter of stripe about one or the other horizontal axis. In Fig. 2.13 we see that symmetry of sorts displayed about the *m* axis.

The rendered graph in Fig. 2.13 is similar to an age-old hand fan. Transverse symmetry is the one-to-one similarity of the slices displayed about the *q*-axis and opposite to one another. This modeling artifact means to express that for one descriptor of a kind, the elements pertaining to opposite slices will behave in some kind of anti-par fashion.

The complete model showcases 4 domains of transverse symmetry as delineated in Fig. 2.14. The complete span of positive values of the function *(+q)* casts four equal quarters, as many as the span of negative values. There is *group symmetry* between the paired quarters such that all the elements of one quarter behave in the same way as opposed to the group of elements in the other matching quarter which behave in a different or opposite way, you might say in an *anti-symmetric* way. Inside of each matching quarter, we have elements in slices or threads that behave in a *transversally symmetric* fashion to one another. These are important descriptors that translate into clear and specific physical properties of matter as we shall see later.

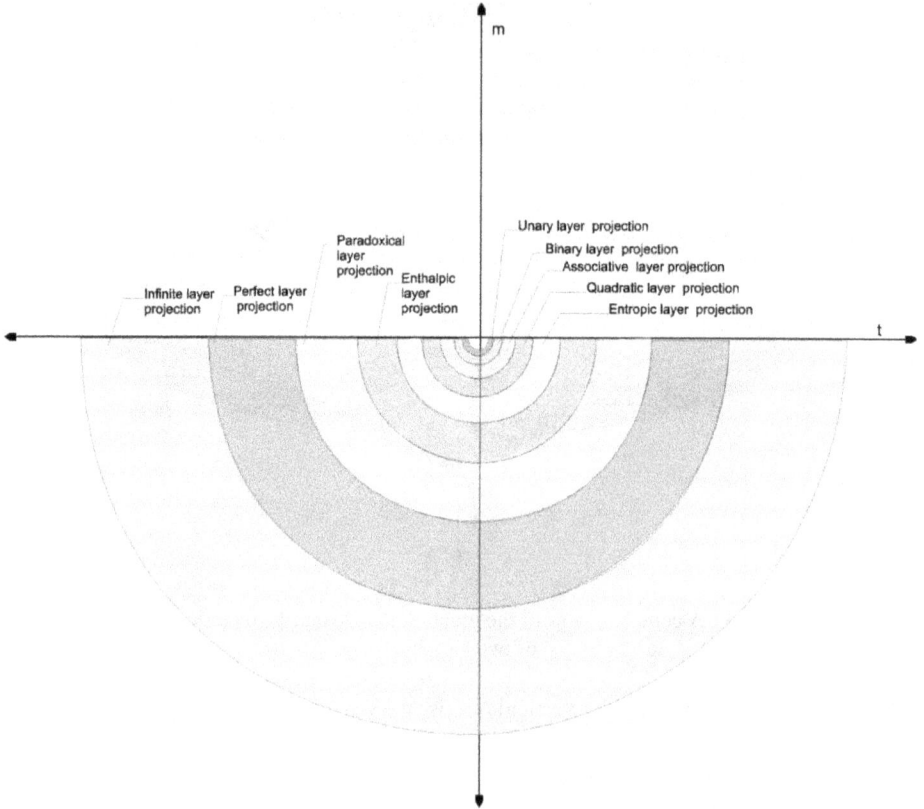

Fig. 2.12 Projection of the three-dimensional Quanto-Geometric graph on the m-t plane. Each washer represents the projection of one Quanto-Geometric layer.

By the same token, elements belonging to the same slice of a thread present among themselves a trait of commonality which we will call *thread symmetry* or *family symmetry*. Those elements are scattered vertically across the 9 normal Quanto-Geometric layers within the thread.

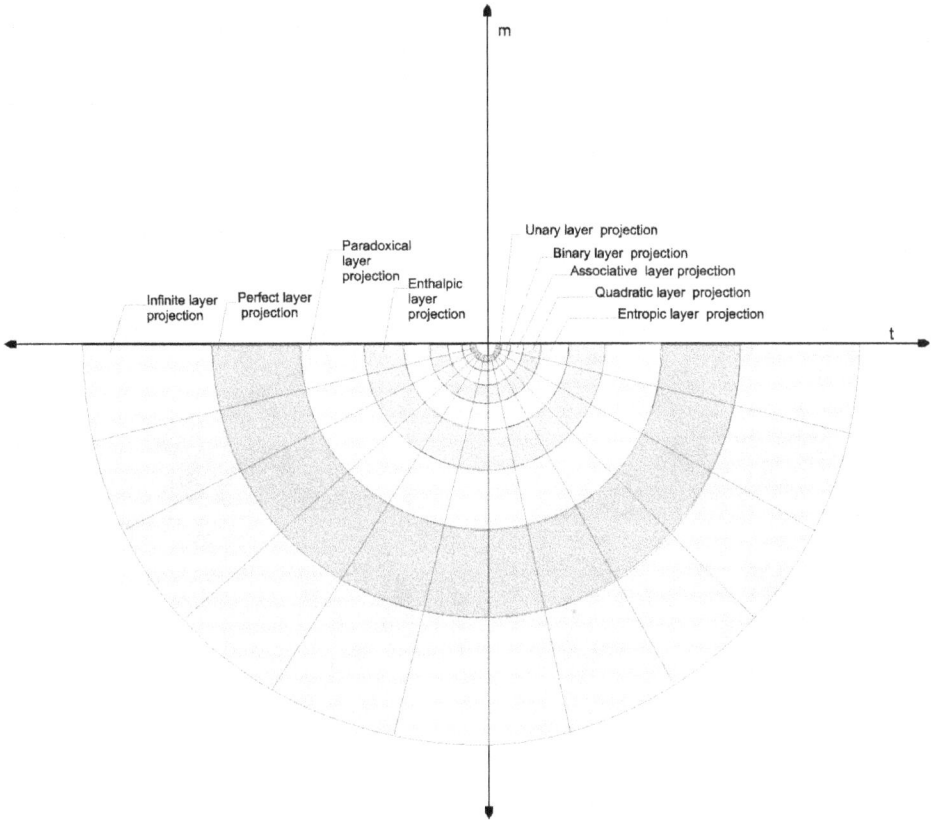

Fig. 2.13 Projection of the three-dimensional Quanto-Geometric graph on the m-t plane with the radii of transverse symmetry about the m axis. Each washer represents the projection of one Quanto-Geometric layer.

To recap, the symmetry groups inherent to the Quanto-Geometric model are:

1. anti-symmetry between pairing quarters

2. transverse symmetry between threads belonging to pairing quarters

3. family symmetry between elements belonging to the same thread

The meaning of the term *symmetry group* is therefore for us very different from that conventionally attributed to it thus far in mathematical physics, in terms of groups of objects experiencing a form of invariance in their physical properties under a particular spatial transformation, rotational or otherwise. In later Chapter we use a specific naming convention to refer to that which otherwise qualifies a usual transformation *Symmetry Group (SU)*.

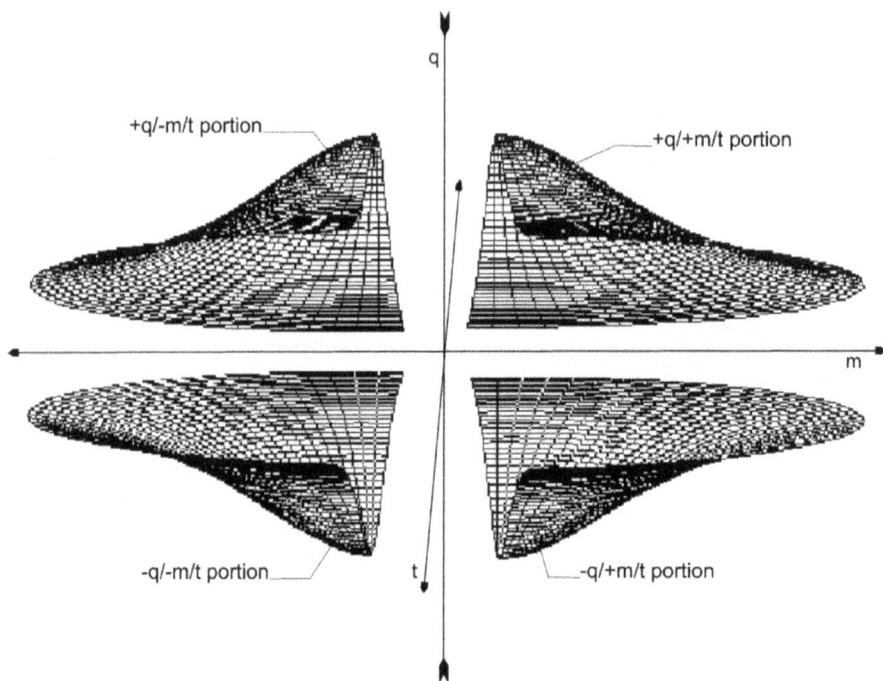

Fig. 2.14 A separate view of all 4 half portions of the graph of *Q(s)*. Each half portion contains two anti-symmetric quarters (symmetry with respect to the *m-axis*).

2.11. Conclusion

In this Chapter we have gone beyond the preliminary principles of Quanto-Geometry and established a formal mathematical formulation of the theory. We have conducted a detailed graphical analysis of the concavity of the Quanto-Geometric function and laid down its properties of symmetry as well as its primal degrees of quantization. We have also looked into one of the 9 Quanto-geometric primitives, the conservation operator, and the correlation with the Gaussian function studied throughout. Finally, we have re-examined the Quanto-Geometric referential and established the conditions it ought to meet in order to respond to the specific requisites of the Quanto-Geometric function.

Chapter 3

Numbers represent the highest degree
of the intelligible.
Plato

QUANTO-GEOMETRIC
NUMBER THEORY

Our primary notion of numbers concerns their cardinality, a quality by which they are applicable to objects, real or virtual, as a way to garner their inventory. And that is what we ask of numbers for the most part. When we state that there are 5 chairs in a room, we are saying that there are chair A, chair B, chair C, chair D and chair E. However, numbers are not just numbers, that is to say quantum descriptors of quantity or amounts.

Every number represents a form of quantity that is not only exogenous to the number as an expression of count applicable to objects in our world, but this form of quantity is equally intrinsic to the number. That intrinsic state of quantity qualifies as a quantum abstraction or an inner state to the number which may be described as a *scalar state*. As such in the numeric domain, there is no continuum, contrarily to what traditional number theory dogma posits. This is true for the numeric domain as a whole (the cardinal domain), including the real, the rational, the irrational, the algebraic and the transcendental ensembles.

It follows from that view that numbers are typed as well. The reason why it is so is ultimately because numbers participate in a Quanto-Geometric spectrum. In all, numbers ought to be visualized as quantum abstractions with specific types in structure and endowed with behavior pattern. This is all to mean that numbers are Operators.

3.12. Number Bases

There are all kinds of number systems in use, from the binary system to the decimal system, and beyond, each established on a base containing a set number of units as building blocks. A number base is whatever you want to make of it in terms of the amount of base units. We are going to utilize the decimal number base (0 to 9), which is familiar to most, in order to undertake our Quanto-Geometric analysis of numeric quantum abstractions.

From the Quanto-Geometric vantage, base numbers, decimal or not, do not exist one after another on a timeline. There are all cast at the same time in a Quanto-Geometric spectrum. By relating the base numbers to one another, we are able to study their identity, their attributes and their behavior, all of which make up their specific Quanto-Geometric signature.

Rules of divisibility are paramount in helping us shed light on the signature of these base numbers. If a number, pertaining to the numeric system but localized beyond the base units, is divisible by a base unit, this is an indication that the base unit or base number represents the building block of that larger number, then designated its multiple. In a sense, if we know the properties of the multiple, we will acquire information about the base number itself, because the properties of the multiple stand in part for a reflection of the intrinsic properties of the base number. That clearly stands to reason since the base number in that case represents the building block of the multiple. Another fruitful exercise is to analyze the results of divisibility of the base numbers among themselves, particularly from 1, in order to establish the primitives for these signatures.

3.13. Decimal Base Number 1

All numbers are divisible by 1. They divide by 1 to remain **integrally** themselves. Thus any number divided by 1 returns whole out of the operation: 1 cannot fraction any number. Number 1 is a monolith whose realm is the Whole. Additionally, all numbers may be built from 1 by **integration** or aggregation. Those characteristics make number 1 a ubiquitous entity and a master key.

3.13.1 Quanto-Geometric Signature:

Identity: Totality – Integrality

Attributes: omnipresence - monolithic

Behavior: As a master key, creates overtures. Centralizes. Aggregates. Individualizes. Conserves. Purports the operation of addition. Begets the ensemble of prime numbers.

3.14. Decimal Base Number 2

The conventional rule states that a number is divisible by base number 2 if it is an even number. However, since the definition of an even number is that it is divisible by 2, the rule is not very much explanatory. An alternative rule is this one: any number ending in 2, 4, 6 and 8 (and 0 as well) is divisible by base number 2. If we restrict the analysis strictly to the base numbers, the statement quickly becomes completely moot as a definition. The following has more of a defining value than the previous: any number that is the sum of two identical numbers is divisible by base number 2. This is expressed as:

$$X = Y + Y \text{ or } X - Y = Y, \text{ in order for X to be even (where X is even)}$$

Hence Base element 2 is a primitive for parity.

3.14.1 Quanto-Geometric Signature:
Identity: Duality

Attributes: Equality – Equivalence

Behavior: polarizes, dichotomizes. Purports the operation of subtraction. Begets the ensemble of even composite numbers.

3.15. Decimal Base Number 3

A number is stated to be divisible by 3 if the sum of its digits is divisible by 3. Once again while this statement can be an expedient rule for dealing with rather small numbers, it bears no merits for a definition since it is calling for a vicious circle of re-divisibility by 3. A better definition is this: a number is divisible by 3 if the ultimate sum or cross sum of its member digits is 3, twice 3 or thrice 3. The ultimate sum means the most non-reducible sum which is obtained after a cascade of summation of the containing digits by resulting numbers if necessary. So therefore divisibility by 3 requires a relationship between the constituent digits of the number. This relationship may be any of 3 different ones and they all involve a direct association with 3. Another way of expressing this condition is that they must all have *one thing in common* as per expression below:

$$ABC = (A+B+C)/3, \text{ where ABC is a multiple of 3.}$$

Base element 3 is a keen proponent of association.

3.15.1 Quanto-Geometric Signature:

Identity: Association

Attributes: Relational

Behavior: Purports the operation of multiplication. Begets the realm of odd composite numbers.

3.16. Decimal Base Number 4

The conventional definition states that a number is divisible by 4 if the last two digits are divisible by 4. Again a statement with little merits for a definition. A better definition yet: a number is divisible by 4 if it can be written as 2^n where n>1. This is a statement of how many equal times the abstract quantum can be fractioned, with the requirement that this number of times be paired. In other words, element 4 **doubles** down on divisibility by requiring a multiplication of the pairing. That is how it introduces the operation of taking the *root* or its inverse, which is raising to a *power*. The number must be of the form:

$$A \times A \times A \times A \times A$$

Base element 4 thus is a proponent of concatenation.

3.16.1 Quanto-Geometric Signature:

Identity: Ambivalence

Attributes: Transitory – Cross-Linking

Behavior: Concatenates. Transitions. Purports the operation of raising to *power* and taking the *root*.

3.17. Decimal Base Number 5

Number 5 poses a rather strange condition of divisibility. Classic number theory has it that a number is divisible by 5 if the last digit of the number is either 0 or 5 itself. That rule has inarguably the perfect value of a definition. The requirement for 0 to be part of the expression of the number is a requirement for the abstract quantum of type 5 to be accompanied by or made up of *nothingness*. It is an invocation of nothingness in the midst of scalar quantity. Furthermore, the requirement that 5 itself be part of the number as an image of itself is an invocation of virtualization. By all accounts the abstract quantum of type 5 is a proponent of the imaginary.

3.17.1 Quanto-Geometric Signature:

Identity: Newness

Attributes: Virtual, metaphoric, metamorphic

Behavior: Transforms. Purports the operation of *complexification* as construed in complex numbers, that is to say adding the imaginary to.

3.18. Decimal Base Number 6

As the rule has it, for a number to be divisible by 6 it has to meet both tests of divisibility by 2 and 3. In other words the abstract quantum of type 6 is calling for both duality and association. Translate *association* by *pair*. This is a requirement for coupling or matching. So therefore the abstract quantum of type 6 is the precursor of all coupling arithmetic, namely matrixial mathematics.

3.18.1 Quanto-Geometric Signature:

Identity: Couple

Attributes: Matchmaker. Orderly.

Behavior: Harmonizes. Purports the operation of *matrification*.

3.19. Decimal Base Number 7

Decimal number 7 is a case of oddities all around. The test for divisibility by 7 rules that all component digits of the testing number be added together, to then test that sum for divisibility by 7, and repeat the whole process until the smallest 2-digit multiple of 7 is possibly reached. Once again this procedure has the value of a test and stands not for a definition. Examining divisibility of base number 1 by 7 gives us a better chance to understand the characteristics of the inherent abstract quantum that makes for its building block. One divided by 7 yields 0.142857. Taking the decimal part of this fractional number two digits at a time, we see 14, 28 and 57. Fourteen is 2 x 7, 28 is 4 x 7. So we see 7 as part of the inner constitution of the number, a paradox that expresses the number's refusal to be fractioned. From that trend, the 3rd set would expectedly be 56, a multiple of 7. However it is 57 instead, in an attempt to showcase very explicitly this time number 7 itself, so portraying the number's determined refusal to be fractioned. In all, all base numbers divided by 7 return the same rotating decimal 142857 as reported below.

1/7 = 0.142857142857142857142857142857 14
2/7 = 0.285714285714285714285714285 7 1429
3/7 = 0.428571428571428571428571428 571 43
4/7 = 0.571428571428571428571428571 42857
5/7 = 0.714285714285714285714285714 28571
6/7 = 0.857142857142857142857142857 14286
7/7 = 1
8/7 = 1.142857142857142857142857142 8571
9/7 = 1.285714285714285714285714285 7143

Note that end of above expressions are rounded off. Many more odd adventures are well known to this number set (142857) in traditional number theory and I could add to the record a few more so far unknown or unpublished. However I will spare this exercise to the reader at this time in the interest of keeping this exposition amenable to all. I will however relate this one, which from the standpoint of our analysis is testimony to the refusal by the number to *construct* or produce straight multiples. If you multiply 142857, which personifies the abstract quantum of type 7, by the units of the number base, the operation returns the same 142857 set in a revolving fashion, except for one case.

142857 x 1 = 142857
142857 x 2 = 285714
142857 x 3 = 428571
142857 x 4 = 571428
142857 x 5 = 714285
142857 x 6 = 857142
142857 x 7 = 999999
142857 x 8 = 1142856
142857 x 9 = 1285713

That result means that quantum abstraction of type 7 cannot *construct* since it fully shows itself back as a result of the construction operation. Thus it is only apt at *de-construction*. In the case of multiplication by 9, you can distinguish among the digital group the sets 28 – 57 – 13 (1), 13 being located at the trailing end and 1 at the beginning. 13 (1) or (13+1) is definitely an attempt to produce the usual 14 which the arithmetic here forbids. In the case of multiplication by 7, because you are already making the intent of multiplying by 7, which is to *de-construct*, the operation is already satisfied. This is argu-ably the reason why it produces the strange result of 999999 which nevertheless lacks multiplicity as a construct.

3.19.1 Quanto-Geometric Signature:

Identity: Paradox, Chaos

Attributes: Odd – Chaotic

Behavior: Operator of Destruction, Antagonism, Conflicts.

3.20. Decimal Base Number 8

Let us directly start by analyzing the results of divisibility of all the base numbers by 8, the conventional rule of divisibility by 8 not helping much our analytical purposes (rule has it that last 3 digits of number must be divisible by 8).

$$1/8 = 0.125 \quad (3-1)$$
$$2/8 = 0.25 \quad (3-2)$$
$$3/8 = 0.375 \quad (3-3)$$
$$4/8 = 0.5 \quad (3-4)$$
$$5/8 = 0.625 \quad (3-5)$$
$$6/8 = 0.75 \quad (3-6)$$
$$7/8 = 0.875 \quad (3-7)$$
$$8/8 = 1 \quad (3-8)$$
$$9/8 = 1.125 \quad (3-9)$$

We may notice at first glance in those results that the digital set 125 is either visibly present in those numbers or portrays itself in a hidden fashion. It is clearly present in *(3-1)* and *(3-9)*. We can see it in *(3-2)* although the number 1 is absent there. We can still see it in *(3-4)* although 1 and 2 are absent. In *(3-5)*, with the number 0.625, it is indirectly present if we realize that 6 x 2 = 12. It is also present in 0.75 *(3-6)* if we realize that 7 + 5 = 12. It is equally present in *(3-7)* with the number 0.875, although deeply hidden here: let's realize that 8 + 7 is 15, which would turn it into 0.155 whereby it could be read as **1** and **2** times **5**. Shall we also say that the figurate 12 is also there under the disguise of 75, which is 7 + 5. We may at least conclude from the above that base number 8 shows a keen awareness for the digital set 125 as if they could be its building block in some fashion. The minimal conclusion at this stage is that **8** is perfectly divisible by the other 8 base numbers in terminating decimals involving the digital set 125.

It is noteworthy that these 3 integers, 1, 2 and 5 are the ones involved in the golden ratio. We remind the reader that the golden ratio is given by:

$$\frac{1}{x} = \frac{x}{1-x}$$

which when developed casts the polynomial: $x^2 + x - 1 = 0$ *(3-10)*

The golden ratio is a perfectly normalized ratio. It has come to embody perfect proportioning or apportionment in applied sciences. The solution to this 2nd degree polynomial is given by the quadratic as:

$$x = \frac{-1 \pm \sqrt{5}}{2}$$

which yields the numbers Φ (phi) and Ψ (psi) where we note once more the digital set 1, 2 and 5 . There is equivalence between the standard golden ratio expression and the known condition of finding 2 numbers whose product and difference are both equal to 1. That condition stands for an *irrationalization* in the real realm. It is expressed as follows:

$$m - n = 1 \quad (3\text{-}11)$$

$$m \times n = 1 \quad (3\text{-}12)$$

From *(3-11)* $m = 1 + n$. If we replace m by this value in *(3-12)*, we obtain:

$$n^2 + n = 1$$

$$\text{and } n^2 + n - 1 = 0 \quad (3\text{-}13)$$

This polynomial is equivalent to *(3-10)*. Consequently both polynomials pose the necessity of two integers engaged in an algebraic relationship. We are going to show that that algebraic relationship is intimately linked to number 8. The condition for resolution of this relationship (the polynomial) is given by the quadratic root. Let's do the following manipulations to the quadratic.

$$x = \frac{-b \pm \sqrt{b^2 - 4ac}}{2a}$$

$$x = \frac{-b \pm \sqrt{b^2 - 4ac}}{2a} \times \frac{4}{4}$$

$$x = \frac{-4b \pm 4\sqrt{b^2 - 4ac}}{8a}$$

$$x = \frac{-4b \pm \sqrt{16(b^2 - 4ac)}}{8a}$$

$$x = \frac{-4b \pm \sqrt{16b^2 - 64ac}}{8a}$$

$$x = \frac{-4b \pm \sqrt{8 \times 2b^2 - 8a \times 8c}}{8a}$$

$$x = \frac{-4b \pm \sqrt{8 \times 2b^2 - 8a \times 8c}}{8a}$$

$$x = \frac{-4b \pm \sqrt{8[2b^2] - [8a][8c]}}{8a}$$

If b is even, that is b = 2d, then

$$x = \frac{-[8d] \pm \sqrt{[8][8d^2] - [8a][8c]}}{8a}$$

Therefore we went from the quest of the nature of the relationship between integers 1, 2 and 5, a number set derived from divisibility by 8, to the quadratic root which incarnates a stunning multiplicity of integer 8. In case you did not notice: 1 + 2 + 5 = 8 ! In conclusion, abstract quantum of type 8 is a promoter of perfectibility and begets the ensemble of algebraic numbers.

3.20.1 Quanto-Geometric Signature:

Identity: Per-fection

Attribute: Irrational

Behavior: Operator of or endows: Completion, Fulfillment.

3.21. Decimal Base Number 9

The rule of divisibility states that a number is divisible by 9 if the sum of its digits is divisible by 9. Since this rule will not help us delve into the ultra-structure of this number, let us start by examining the results of divisibility of the base numbers by 9.

```
1/9 = 0.111111111111111111111111111111111111111111111111111111
2/9 = 0.222222222222222222222222222222222222222222222222222222
3/9 = 0.333333333333333333333333333333333333333333333333333333
4/9 = 0.444444444444444444444444444444444444444444444444444444
5/9 = 0.555555555555555555555555555555555555555555555555555555
6/9 = 0.666666666666666666666666666666666666666666666666666666
7/9 = 0.777777777777777777777777777777777777777777777777777777
8/9 = 0.888888888888888888888888888888888888888888888888888888
9/9 = 000000000000000000000000000000000000000000000000000000001
```

All base numbers divided by 9 return an endless suite of digits of the same dividing base number, demonstrating an affinity for serialization. By the same token, it so demonstrates the greatest level of linearity among all base numbers, since the series are made up of the same numbers. One may think of linearity as the inverse of multiplicity, here at its minimal expression.

The serial sum of all base numbers returns an image of 9 as shown below.

$$1 + 2 + 3 + 4 + 5 + 6 + 7 + 8 + 9 = 45$$

which yields a multiple of 9 and the cross sum $4 + 5 = 9$. Likewise if we multiply all base numbers together in a string, we obtain an image of 9.

$$1 \times 2 \times 3 \times 4 \times 5 \times 6 \times 7 \times 8 \times 9 = 362880.$$

The image resides in the fact that the cross sum:

$$3 + 6 + 2 + 8 + 8 + 0 = 27$$

which yields the other cross sum $2 + 7 = 9$. In addition, the number 362880 is a multiple of 9. Now of course the digital series $1 \times 2 \times 3 \times 4 \times 5 \times 6 \times 7 \times 8 \times 9$ is **9!** (9 factorial). Let us remind the reader that a factorial is defined as a downward serial product starting with the designated number.

Hence we shall define divisibility by 9 as the ability for any number to meet the following test: may be written as: **n!** for all n ≥ 6. Hence factorialization, a form of serialization, becomes an important expression of abstract quantum of type 9.

Let us now consider the following series, known as a series of the power type, that is to say a power series:

$$\sum_{n=0}^{\infty} \frac{x^n}{n!}$$

This is the expression of elevating a number x to a power n, divide that power number x^n by the power factorial $n!$, and finally taking the serial sum of all the terms starting with power n equals 0. Clearly, even though the number that we are acting on is a power, we are primarily serializing that number, because firstly we are adding a suite of terms in sequence with one another, and secondly we are additionally introducing another layer of serialization as well with the division of the number by a sequential factorial. The developed expression of the series commonly reads like the following:

$$\sum_{n=0}^{\infty} \frac{x^n}{n!} = 1 + x + \frac{x^2}{2!} + \frac{x^3}{3!} + \dots$$

We are going to rewrite the right side of the expression to make sure it is understood as a thorough suite of progressing terms.

$$\sum_{n=0}^{\infty} \frac{x^n}{n!} = \frac{x^0}{0!} + \frac{x^1}{1!} + \frac{x^2}{2!} + \frac{x^3}{3!} + \dots$$

Any number raised to 0 power is equal to 1 ($x^0 = 1$) and 0! is defined equal to 1 by convention. We end up with 1/1 which is equal to 1 for the first term of the series. As to the second term, since 1! is straightforwardly equal to 1, the second term is clearly equal to x. This series is an infinite sum, we can push it to the largest n integer that we are possibly able to conceive. The numerator of our fractions are serializing as much as the denominators are. The denominators are furthermore factorialized. It is not so much about the powers at the numerators but the fact that they are serializing. Now, this infinite series, sum of an infinite suite of serialized terms, converges to one number, which is *transcendental e*. Effectively, this series is equal to e^x as stated below, no matter how far we push n.

$$\sum_{n=0}^{\infty} \frac{x^n}{n!} = e^x$$

So therefore the operation of serialization bears the essential potential of creating transcendental numbers. Those are the inner workings of abstract quantum of type 9.

3.21.1 Quanto-Geometric Signature:

Identity: Endlessness

Attribute: Serial. Transcendental.

Behavior: Operator of Infinition or Asymptotization.

3.22. Quintessential Geometric Order

Akin to the scalar order that we have just studied, there exists a space or vacuum state order as well. Although the nature of physical space is distinct from mind, absolute pure space, which does not physically exist, can be perfectly assimilated to the abstraction of nothingness. One may wonder how can an order of things in state of nothingness even be formulated? Truth is that the ultra-structure of state of nothingness can indeed become tangible thru quintessential geometry or the order of forms, just like scalar order may be elicited by abstract quanta representing its building blocks per previous discussion.

A simple thought exercise may help us initiate that analysis. We choose the Euclidean plane to develop our abstraction. What is the simplest thing we can do on that plane to start out? As a minimum we can draw a line segment. Not a rectangle (defined by two line segments) as contented in traditional formalism. Not even a dot, because a dot is a concentric element, hence a scalar entity. If you insistently contend that it must be a dot, then conceded, but it must be two dots, the reason being that our Euclidean plane only recognizes linear elements, and two dots are very much a condition for a line. Consequently we initialize our abstraction with an A line as our element 1. This line is Unique, Total and Autonomous.

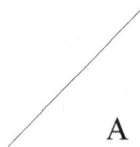

A

As long as we remain on that line, no further geometrization is possible since the line is limitless on both ends. We must step aside from the line in order to strike the next geometric degree. The least subsequent operation is to draw a second line. It can either be parallel to the first or cross it thru. If a parallel line, we may liken this operation to a symmetric movement which amounts to a repetition. In that event there is no new operation. Therefore our second line must necessarily cross the first. No matter the orientation we give the second line, it will always be interchangeable with the first one. Consequently our B line materializes the abstraction of Similitude.

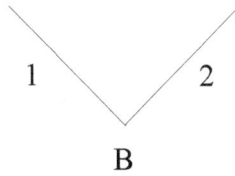

B

Next step in our movement leads us to leave the second line in order to draw a third one. It will mandatorily cross the first two, either in an irregular fashion, or in a regular fashion by forming an isosceles or even equilateral triangle. Whatever the case, the 3rd line establishes a reunion of the first two, therefore positing the axiom of Association or Reunion.

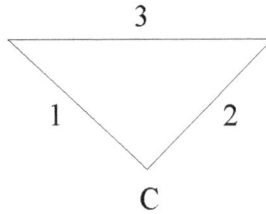

C

In continuing this abstract exercise, let us suppose that the triangle resulting from the above movement is regular, isosceles for instance. The triangle figure, whether regular or not, represents a closed space eliciting completion. Hence, the next geometric step must occur externally to the triangle.

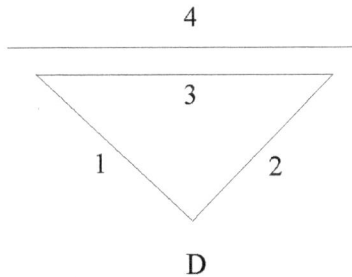

D

Our next line must be drawn with no interaction with the triangle. This obligated rupture and absence of common measure with the fore places the fourth line in the isolation of an end. Thus it must be drawn in parallel with the third line. However this end can only be Transitory since we may continue with the movement.

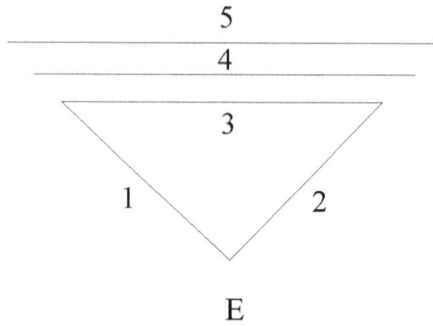

E

The rupture and isolation posited by the fourth geometric unit lends to the next one an origin and differentiation value from which we may reinitiate our geometric movement. Therefore the fifth must be drawn beyond the fourth and must remain parallel to it in order to avoid a violation of the Rupture value incarnated by the fourth. Being a representative of origin and differentiation, the fifth ought to embody Novelty, Change or Transformation.

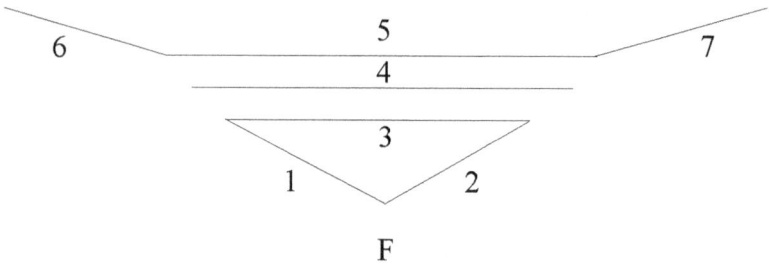

F

What are the criteria that may allow us to continue with our abstract exercise? We know two lines were necessary in order to make up the prime angle (angle formed by A and B). We may now emulate the primordial geometry by drawing two lines in an angular fashion from the fifth, as though they originated from the virtual continuation of side 1 and 2 of the prime triangle. In doing so, we remain in linkage with the primordial geometry while striking a new (post-geometric) operation. However we must obligatorily introduce a distinction of sorts to this new geometric step: we will have for such the sixth and the seventh drawn per a divergent angle from the primordial angle as shown in Fig F.

The sixth and the seventh lines gather these notions and become the expression of Concordance and Divergence, the first characterization being applicable to the sixth geometric unit and the second to the seventh unit.

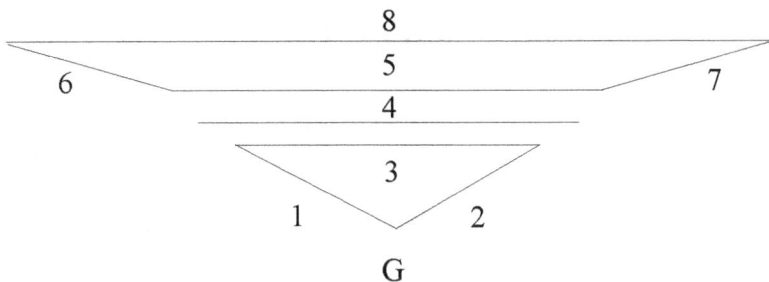

G

We may continue the exercise by uniting the sixth and the seventh with an eighth line. The eighth so confers to the post-geometric movement its utmost completion. We may say that the eighth affords its highest degree of wholeness to the post-geometric movement by configuring a closed trapezoid. It grants the post-geometric movement with its highest level of concretion and becomes the representative of Completion and Per-fection (see Fig G).

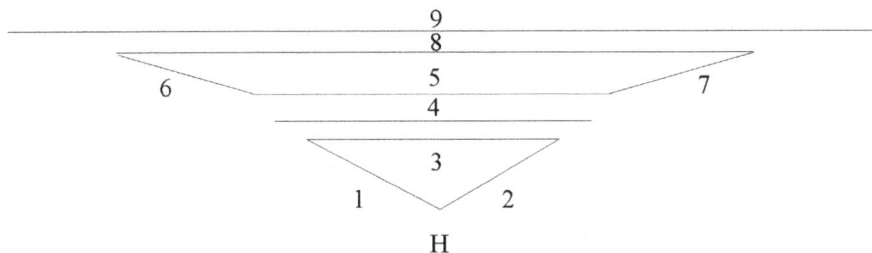

H

Being complete in and of itself, the eighth permits no more crosses. Its wholeness forces the ninth to remain independent by parallelism, in a movement that is akin to the fourth. However, after the most Perfect Completion, only an absolute ending may occur, that is to say an end from which nothing else can originate, lest we repeat all previous operations. Consequently, the ninth rejects all crosses as well as stops. It extends in one or the other direction indefinitely and unconditionally. That is how the paradox of endless end comes about: the Asymptotic End.

This abstraction which we conducted on the basis of a Euclidian plane may be realized as well on a tri-dimensional reference system with the exact results and conclusions. There undoubtedly exists an ultra-structure within the state of nothingness based on 9 geometric primitives in the terms that we have exposed above. The ensuing geometry elicits in forms these inner states of the order of nothingness. One can further see that these inner states are tightly matched with the abstract quanta of the scalar order. More on that correspondence later.

3.23. States of Symmetry in Nothingness

The inner states of the order of nothingness are states of symmetry. It is important for us to recognize that forms of one kind or another always amount to degrees of symmetry. A geometric form is a *form* of *symmetry*. The reason why things get to hold a form is because of the existence or primacy of these abstract states of symmetry which make up the order of nothingness. Just like every object automatically has a count which springs from the scalar order, every object automatically has a form that is rooted on the inner states of symmetry pertaining to the order of nothingness germane to the object.

It is by articulation with quanta of the scalar order that the states of symmetry pertaining to the order of nothingness come to tangibility under the wrapper of forms (Fig. 3.1). Geometry is a materialization of these forms in the physical world, it is a *formal* representation. Later on we will further discuss the terms of this Quanto-Geometric articulation.

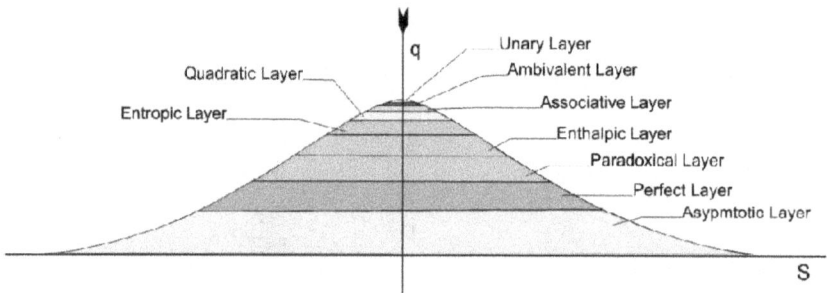

Fig. 3.1 The nine primal Quanto-Geometric types represented
by the intrinsic layers of the $Q(s)$ function

Notwithstanding their abstract nature, it is not possible to completely separate from one another the two primal orders, the scalar order and the order of nothingness. Therefore geometric formation appears within the ultra-structure of numbers, as much as scalar behavior is perceptible within the order of nothingness per our former exercise of geometrization. We can *count*, I repeat *count*, the geometric states and this is not a trivial statement. This means that they have a *cardinal* reality which is mapped to the scalar units of the scalar order in a one-to-one relationship. You may as well *ordinalize* the geometric states, that is to say establish a progressive rank among them. Traditional Number Theory professes that cardinalization and ordinalization are the two foremost qualities of the number domain.

The most important consequence from the above proposition to bear in mind is that the qualities that we have derived for each state of symmetry match the qualities that we have derived for scalar units from the scalar order. This is because there is a tight Quanto-

Geometric articulation between the two orders, whereby they both influence each other. It is the influence exerted on the order of symmetry by the scalar order which produces its cardinalization and ordinalization. Conversely the number domain presents geometric properties as a result of influence exerted by the order of nothingness or symmetry. Shall we take a look at these geometric properties elicited by the scalar order.

3.24. Geometric Properties of the Scalar Order

It is well-known that real numbers show a handful of geometric properties. For instance real numbers characterized as *palindromic* are numbers that read the same forward or backward. For instance 12721 is palindromic. We can easily see that a palindromic number must have an odd number of digits so that the mid integer may stand as its center of symmetry. If the number is turned around about that mid digit as its center of symmetry, it reads exactly the same. We can also invoke *figurate* numbers, which are numbers that are structured according to a geometric pattern. For instance, pentagonal numbers, sexagonal number, heptagonal numbers and octogonal numbers are all numeric entities well described in the literature (Fig. 3.2).

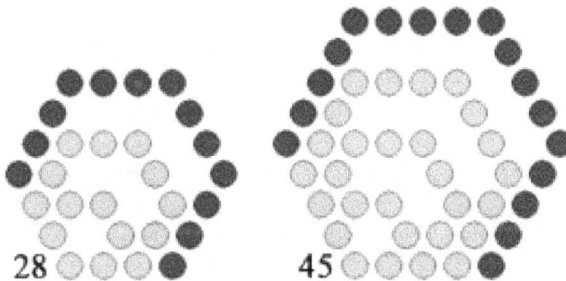

Fig. 3.2 Polygonal Numbers

Our point of interest herein however is to show the geometric properties evidenced by the intrinsic scalar order itself. We are going to study the nature and distribution of numbers in the real domain in order to develop this thesis. The dogmatic approach followed in the classroom and in text in presenting mathematics in general tends to make us lose sight that this science is all about the nature and distribution of numbers.

0	1	2	3	4	5	6	7	8	9	10
11	12	13	14	15	16	17	18	19	20	21
22	23	24	25	26	27	28	29	30	31	32
33	34	35	36	37	38	39	40	41	42	43
44	45	46	47	48	49	50	51	52	53	54
55	56	57	58	59	60	61	62	63	64	65
66	67	68	69	70	71	72	73	74	75	76
77	78	79	80	81	82	83	84	85	86	87
88	89	90	91	92	93	94	95	96	97	98
99	100	101	102	103	104	105	106	107	108	109
110	111	112	113	114	115	116	117	118	119	120
121	122	123	124	125	126	127	128	129	130	131
132	133	134	135	136	137	138	139	140	141	142
143	144	145	146	147	148	149	150	151	152	153
154	155	156	157	158	159	160	161	162	163	164
165	166	167	168	169	170	171	172	173	174	175
176	177	178	179	180	181	182	183	184	185	186
187	188	189	190	191	192	193	194	195	196	197
198	199	200	201	202	203	204	205	206	207	208
209	210	211	212	213	214	215	216	217	218	219
220	221	222	223	224	225	226	227	228	229	230
231	232	233	234	235	236	237	238	239	240	241
242	243	244	245	246	247	248	249	250	251	252
253	254	255	256	257	258	259	260	261	262	263
264	265	266	267	268	269	270	271	272	273	274
275	276	277	278	279	280	281	282	283	284	285
286	287	288	289	290	291	292	293	294	295	296
297	298	299	300	301	302	303	304	305	306	307
308	309	310	311	312	313	314	315	316	317	318
319	320	321	322	323	324	325	326	327	328	329
330	331	332	333	334	335	336	337	338	339	340
341	342	343	344	345	346	347	348	349	350	351
352	353	354	355	356	357	358	359	360	361	362
363	364	365	366	367	368	369	370	371	372	373
374	375	376	377	378	379	380	381	382	383	384
385	386	387	388	389	390	391	392	393	394	395
396	397	398	399	400	401	402	403	404	405	406
407	408	409	410	411	412	413	414	415	416	417
418	419	420	421	422	423	424	425	426	427	428
429	430	431	432	433	434	435	436	437	438	439
440	441	442	443	444	445	446	447	448	449	450
451	452	453	454	455	456	457	458	459	460	461
462	463	464	465	466	467	468	469	470	471	472
473	474	475	476	477	478	479	480	481	482	483
484	485	486	487	488	489	490	491	492	493	494
495	496	497	498	499	500	501	502	503	504	505

Fig. 3.3 Status of the first 505 positive integers on the real number line: 96 primes, 155 odd composites and 254 even composites

It is almost always about finding a number or set of numbers which satisfy a certain condition, that condition being that which is expressed in a function, a matrix, a series or other manners of mathematical expressions presented upfront.

Two large classes of numbers are known to make up the real number domain: integers and decimals. These two classes are distributed in layers which we shall review shortly.

1.- At the top of this hierarchy are the *Prime* numbers. Primes are numbers that can only be divided by 1 and themselves. They are indivisible monoliths. Although there are an infinite number of primes, there are less primes than any other types of numbers in the whole number realm. The table in Fig. 3.3 above shows the primes among the first 505 real numbers (those in yellow or light gray cells) of the number line. There are only 96 primes among the first 505 real numbers.

2.- The next layer is that of composite numbers, specifically odd composite numbers. Hence at layer 2, we are first of all dealing with *composition*, that is to say elements made up by two or more elements. *Composite* numbers are multiples of other numbers. Secondly, one may wonder why we are restricting this class to odd numbers. So we do because we are executing the inverse of a densification function, we are breaking down the monoliths from layer 1. All composite numbers, including odd composite numbers, are the result of multiplying primes together.

Because composite numbers result from combination of primes, there will be more composite numbers than the monolithic primes. There will even be more odd composite numbers than primes.

3.- Layer 3 represents the class of even composite numbers, which are numbers divisible by 2. This class will have more numbers than the previous, the set of odd composite, for 3 main reasons:

1. Element 2 of the base despite being a prime is an even number (the only even prime); as a muliplicand (factor) it will make every number it generates an even number.

2. Element 0 or 10 of the base is an even number and as a muliplicand it will make every number it creates an even number.

3. While every even number multiplied by an even number creates another even number and every odd number multiplied by an even number creates an even number, every odd number multiplied by an odd number creates another odd number however.

Therefore the class of even composite will be a lot more prolific than the class of odd composite.

4.- Layer 4 showcases root or power numbers. These are a sort of special composite numbers which are the result of identical multiplicands. For instance $5 \times 5 = 25 = 5^2$, so 25 is a power number of which 5 is the root. Another example: $6 \times 6 \times 6 = 216 = 6^3$, so therefore 216 is a power number of which 6 is the root.

Is this class more prolific than the previous? It sure is and the reason is simple: since any integer, not only the base elements (2 to 9), may be multiplied by itself 2, 3, 4, 5... a limitless number of times, to produce power numbers, then from only one strain of power (i.e. $8 \times 8 \times 8$ ...) you are already generating a divergent series of numbers toward infinite. If the previous classes were open sets toward infinite from one, two or three streams, the class of power numbers opens up to infinite from a larger and larger number of streams that are themselves limitless. This characterization brings the cardinality of this class to experience an explosion of sorts which places it much beyond the cardinality of the previous class. By all accounts this behavior represents an inflexion point in the cardinality of the vertical hierarchy of classes herein under scrutiny.

5.- Layer 5 is the province of complex numbers. These are numbers made up by a *real* part added to an imaginary part, that imaginary part being made up of a coefficient of i (i is equal to $\sqrt{-1}$) as in *6i* or *25i*. The real part can be an integer belonging to any one of the previous classes. On that count alone, this class would surpass the previous one in cardinality. But when you add the imaginary part, the coefficient of which may belong to any of the previous classes as well, then we can see that the amount of possible combinations between the real part and the imaginary part puts us well beyond the count of the previous class at layer 4.

6.- Layer 6 showcases the class of rational numbers or decimals resulting from ratios, specifically those numbers with terminating decimals (i.e. 6.5) and repeating decimals (0.166666666666666). One can easily appreciate that the variety of decimals that may be achieved in these two manners can give way to a very high level of cardinality that surpasses the cardinality of layer 5. Let us first realize that the domain of integers (elements from layer 1 to 4) can take the place of the whole part in these decimal numbers. Secondly, to each of these whole integers, we may add an infinite number of terminating decimals of different lengths and an infinite number of repeating decimals of different lengths as well. And obviously if we wanted to turn these numbers into complex numbers, we could add these same sets to the imaginary part of these numbers as coefficients of i.

7.- Layer 7 is the province of irrational numbers, which are decimals with non terminating, non repeating decimal portions endowed with no particular patterns. They are not the result of integer ratios. The most surprising account about irrational numbers is that there are more of them than rational numbers on the real number line, as proven by Danish-German mathematician Georg Cantor in the 19th century. Common sense then and still today would tend to make us believe that there would have to be more rational numbers than irrational numbers in existence. It is shockingly and strictly not so however as proven by Cantor.

8.- Layer 8 represents the class of algebraic numbers. These numbers are defined by clear geometrical relationships as we have shown in the study of the order of nothingness. It is well established that their cardinality is higher than that of irrational numbers on the real number line.

9.- Finally Layer 9 portrays the class of transcendental numbers. These numbers are defined as well by geometric relationships. The most famous among them, π and e, are respectively defined by relationships between elements of a circle and elements of a function ($1/x$). It is equally well established that among irrational numbers (those starting from layer 7) the set of transcendental numbers have the highest cardinality.

Despite the fact that all 9 classes of numbers on the number line form infinite sets, the cardinality of each set augments from class to class down the vertical hierarchy, with the important fact that class 4 shows an explosion in cardinality over the previous class (class 3). That explosion we have identified as an inflexion in the numerical analysis above. Therefore there is a clear geometric pattern overall to the distribution of numbers in the realm of real numbers. You may even liken it, if you will, to the distribution figure shown below in Fig. 3.4.

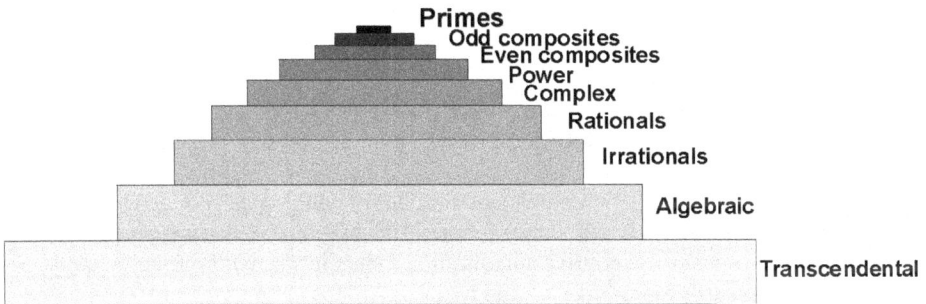

Fig. 3.4 Real Number Stack

At any rate, this pattern is nothing else than the Quanto-Geometric pattern or the bell-shaped pattern (see Fig.3.5). The inflexion point at layer 4 is indeed a standard deviation point in nature. What causes the various sets of numbers to behave in that manner is the intrinsic state of symmetry embedded in the constitution of these numbers within their sets. In other words, it is the reflection or influence of the order of symmetry over the scalar order within the Quanto-Geometric articulation which causes the scalar order to behave geometrically. It is not quite possible to plot a graph of the scalar distribution of real numbers that would capture the fullness of the number stack, simply because each layer takes origin from $-\infty$ and tends toward $+\infty$. Nonetheless it remains utterly clear that the concept of the number line is far insufficient and probably inadequate to capture the ultra-structure of real number elements.

By the same token, the geometric distribution of real numbers per above analysis has an overwhelmingly important consequence: it reveals that *the number line is not a continuum, the number line is discontinuous*. Effectively the distribution of real numbers studied above represents the highest level of abstraction possible if one is to appropriately model the domain of real numbers. This abstraction represents a meta-model of the overarching structure and behavior of real numbers. The ensuing distribution cannot be plotted on any of known coordinate systems, perhaps only the Quanto-Geometric one set forth in this study, because it is utterly difficult if inappropriate to use numbers to model numbers. Our best attempt at a pictorial representation of the distribution spectrum of the realm of real numbers is the graph shown in Fig. 3.5.

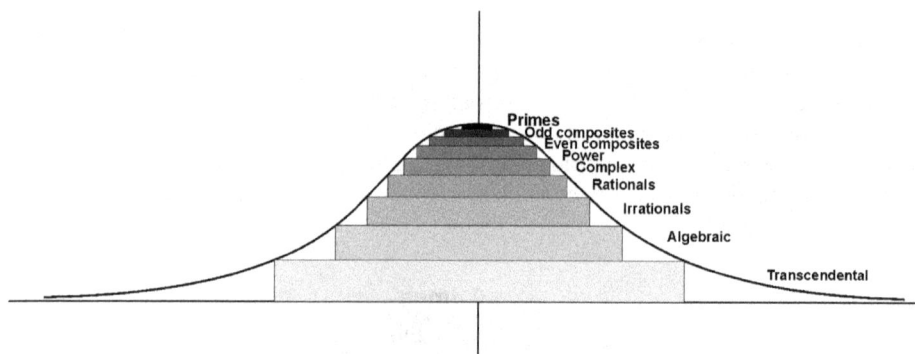

Fig. 3.5 Quanto-Geometric envelope of the real number stack

The horizontal axis stands for an index of cardinality which speaks of generic orders of magnitude while the vertical axis stands for an index of global scalar weight (or scalar modulus) so as to image the relative amount of numbers in sets with respect to unity. Unity in this context is represented by the set of monoliths (Primes). That approach makes the vertical axis a modeler for density figures. On the basis of that coordinate system the distribution graph naturally follows as depicted in our representation. Thus the monoliths represented by Primes therein have the highest index of scalar weight but the lowest count or index of magnitude, whereas transcendental numbers hold the lowest index of scalar weight but highest count or index of cardinality (the magnitude). The graph shows an inflection point and reveals itself symmetric to the Quanto-Geometric function. Again, more than a functional representation, this characterization stands rather for the depiction of a graphical envelope to the real number stack (Fig. 3.5).

3.25. States of Symmetry in the Quanto-Geometric Function

In closing this chapter it is worth revisiting the Quanto-Geometric function with the aims of further eliciting the inner articulation between scalar states and states of symmetry. That function is based on transcendental numbers such as e, π, and $\sqrt{2}$, located at the highest level of the real number distribution, thereby making it apt to describe developments at all the other layers of the distribution.

I cannot overstate at the outset that objects as well as phenomena in physical reality are **numbers's applications**. That is to say that all things physical do not only exist thru number cardinality (their intrinsic numeric count) but also thru the scalar state and state of symmetry that make up their cardinality. On that account, every physical object is typed by the scalar and symmetric qualities pertaining to their cardinality, which endow them with specific properties emanating from the inner numeric ultra-structure. Those qualities and properties are the *Quanto-Geometric* descriptors set forth in previous chapters.

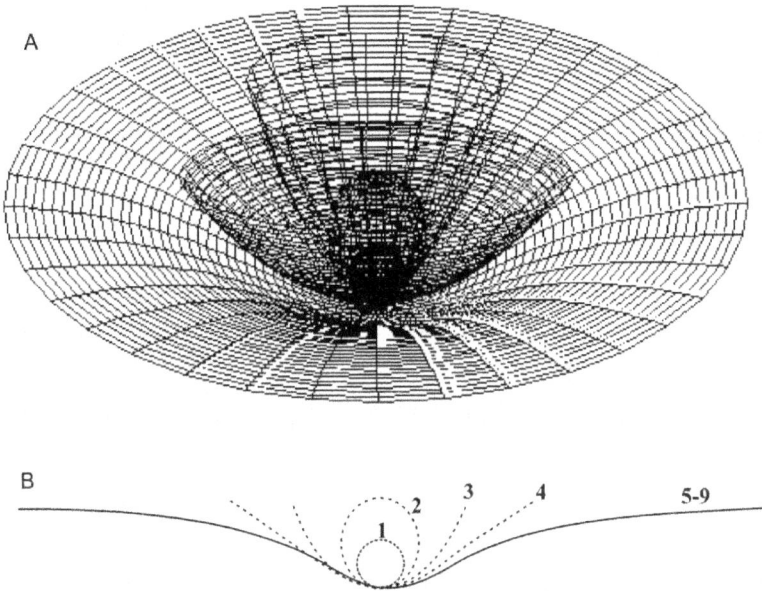

Fig. 3.6 Delimitation and typification of normalized intervals of the Quanto-Geometric function.

We are going to delve somewhat deeper into the correlation between scalar states and symmetric states from our abstract framework as reflected in the Quanto-Geometric function. We are going to revisit the *s* intervals of the Quanto-Geometric function and the curvature of the function across those intervals (Fig. 3.6).

One may easily observe a distinct form of curvature of the function for each interval (Fig. 3.7). At s = 0, the Quanto-Geometric function approaches a limit. In the interval $]0, \sigma/2[$, the curvature of the graph is circular. A circle, having only one focus which is its point of origin, translates into all points of the circle having only one way to compare to or resemble with one another. That one center of symmetry is a **singular** state of symmetry which matches the cardinality of that layer: numeral 1. In other words scalar state 1 correlates with state of symmetry 1 at layer 1.

In the next interval $[\sigma/2, 2\sigma/7^{1/2}[$, the geometry of the curvature is elliptical. Since an ellipse is based on two foci, one may assert that there are two ways that the points of an ellipse may compare to or resemble with each other. This also means that there is two times more symmetry in the ellipse figure than there is in the circle. Thus the ellipse contains a **dual** state of symmetry which matches the cardinality of that layer: numeral 2. Thus scalar state 2 correlates with state of symmetry 2.

The section of curvature laying within the interval $[2\sigma/7^{1/2}, \sigma[$ is parabolic. Since a parabola in defined by one line, the directrix, and one focal point, the both of which are reducible to 3 points in total, one may conclude that the points of a parabola may compare to or resemble with each other in three different ways. This also means that there is three times more symmetry in the parabola than there is in the circle. Therefore the parabola is endowed with a **triple** state of symmetry which matches the cardinality of that layer: numeral 3. Thus scalar state 3 correlates with state of symmetry 3.

In the interval $[\sigma, \sigma(8/5)^{1/2}[$ the function is of hyperbolic curvature. Since the geometry of the hyperbola is based on 4 centers of symmetry, the rectangle that defines the two axes of representation thru its diagonals, one may affirm that the hyperbola possesses a **quaternary** state of symmetry which matches the cardinality of that layer: numeral 4. Consequently scalar state 4 correlates with state of symmetry 4. The forms of curvatures we know as conic sections end up with this interval.

The next five s intervals delineate forms of curvature known as hyperbolic from the hyperbolic functions, specifically the *sech* function. We have shown in Chapter 1 how the expression in e of the Quanto-Geometric function may be rewritten in implicit form to show the expression of the hyperbolic *sech* function. Assigning different values to the variable a, b and c in the implicit expression will yield graphical curves falling into interval 5, interval 6, interval 7, interval 8 and interval 9 of the Function per our form of normalization.

In the fifth s interval $[\sigma(8/5)^{1/2}, \sigma5^{1/2}[$ the function displays a form of curvature we call the entropic form of curvature, which stands for a state of symmetry based on 5 centers of symmetry. Consequently state of symmetry 5 correlates and articulates with numeral 5 or scalar state 5.

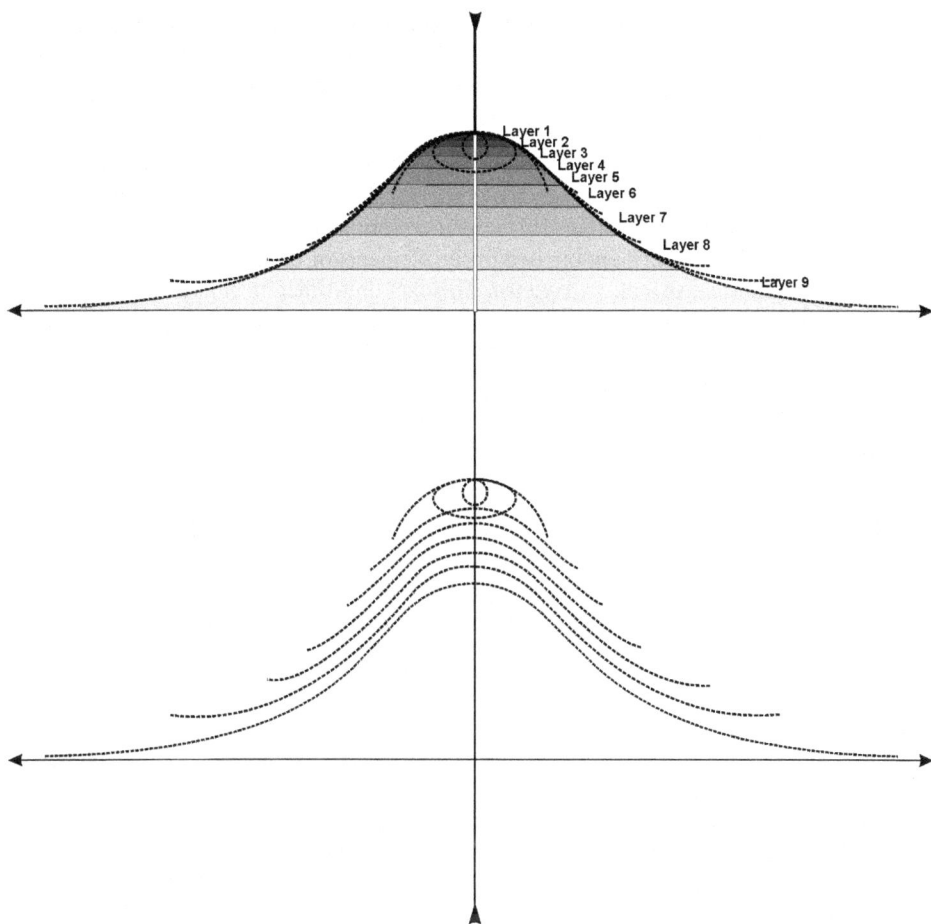

Fig. 3.7 Exploded view of embedded symmetries within the Quanto-Geometric function

The sixth interval $[\sigma 10^{1/2}, 2\sigma[$ delineates a form of curvature we denote the enthalpic curvature, which stands for a state of symmetry based on 6 centers of symmetry. Thus state of symmetry 6 correlates and articulates with scalar state 6.

In the seventh interval $[2\sigma, \sigma 7^{1/2}[$ the function describes a form of curvature that we call paradoxical curvature, which stands for a state of symmetry based on 7 centers of symmetry. Thus state of symmetry 7 is in articulation with scalar state 7.

The eighth interval $[\sigma 7^{1/2}, 4\sigma[$ encompasses a fraction of curvature which we call the perfect curvature. This state of symmetry is based on 8 centers of symmetry. Thus state of symmetry 8 is in articulation with scalar state 8.

And finally in the ninth interval [4σ, ∞[the function displays a section of curvature we denote linear or infinite curvature, which stands for a state of symmetry based on 9 centers of symmetry. Thus state of symmetry 9 is in articulation with scalar state 9.

What justifies our assigning an attribute to those states of symmetry is the correspondence that we have observed between these attributes when determined by either the geometric order or the scalar order. It is now therefore firmly established that the Quanto-Geometric articulation, between the two prime elements of the ultra-structure and across the abstract numeric distribution, is *typed*. This fact or artifact indeed raises each layer of the distribution to the value of an Operator when the number system is applied to physical reality, or, do we dare say, when the number system is put to work to create physical reality. Is this a reflection of the Mind of God at Creation? I let you make that call.

3.26. Conclusion

The numerical analysis that we have conducted above is based per choice on the decimal number system with 10 base units. The identity of base unit 0 has been in dispute for quite a while now, some mathematicians having removed it altogether from the number line as origin. The prevalent supporting argument has always been that *no-thing* cannot belong to a set of *some-things*.

The best way to formulate an abstraction about 0 is thru the philosophical principles that we have proposed in Quanto-Geometry, as expressed in Chapter 1. There is no absolute 0 in physical reality. There is only absolute 0 in complete and total abstraction, and that is why we have subsumed that 0 into the coordinate system's *m-t* plane or the *s* axis which acts as an unreachable asymptote. When the abstraction comes down to the number realm, 0 starts having some life or becomes part of that abstract life as an order of nothingness.

It is because of that 0 that numbers of the scalar order behaves the way they do in terms of their geometric properties and their distribution. Even if we were to remove 0 from the set of base units, making it a base of 9 units, the properties and the distribution would remain the same, because there cannot be any scalar order without an order of nothingness. From the Quanto-Geometric perspective, the base of 9 units (1 to 9) is the most natural of base number systems. However there is no prejudice whatsoever in keeping 0 within a number base, as does the decimal base, because the meta-level abstraction that presides on all number systems does not formally forbid its inclusion. I shall equally add that the conclusions and results that this numerical analysis has reached would not have changed had we used a different number system than the decimal.

By and large, all the facts known in Number Theory, particularly those known in *recreational* number theory, are relevant facts within the context of the Quanto-Geometric vision. Thus far the many facts known about a long string of specific numbers which have received rather exotic attributes and have been very loosely categorized, reason for their denomination as *recreational*, will find fully coherent and systemic interpretation within the Quanto-Geometric framework. We are refraining ourselves from further exposition in that direction in the interest of not overwhelming readers with possibly little interest in this topic, no matter how recreational mathematicians think this section of Number Theory is. Nevertheless it must remain absolutely clear that numeric abstraction is the foundation of all physical reality as they appear under the umbrella of both objects and phenomena. The most towering vision that should remain stamped in the reader's mind is the artifact that the ultra-structural signatures of base number units of the real number realm are the source from which spring the behavior, attributes and identity of all real physical objects and phenomena. This fact or artifact has been altogether ignored in the West despite the 2000-year-old warning from the brilliant Greek philosopher Aristocles, nicknamed **Plato**, that *numbers represent the highest level of the intelligible.*

Chapter 4

Everything should be as simple as possible,
but not simpler.
A. Einstein

QUANTO-GEOMETRIC TENSORS
AND OPERATORS

Physicists have long realized that the idea of holding one theory for the treatment of dynamic systems in the microcosm and one other for the dynamics taking place at larger scale physical structures is simply untenable, moreover if both families of theories are completely opaque to one another. Despite the proliferation of gauge theories, we are still unable to explain the continuum of Nature as revealed before us and it may be time to realize that the bridging notion of quantum gravity is just ill-founded and non-falsifiable as framed. There will be no Grand Unification if we do not endeavor to elucidate the origin of the most important traits of the continuum, which is the plethora of fundamental physical constants that make up the continuum of Nature, those which we have arrived at by pure empiricism for the most part. In this Chapter, I lay down the conceptual framework that we have dubbed **Quanto-Geometry** in order to address the ontological nature of space and the quantum-scalar along with their correlation throughout the continuum, in addition to the pivotal Tensors and Operators acting on the continuum.

4.1. Duality and Codependency in Quanto-Geometry

To achieve the afore-mentioned goals, we recognize that the Theory must be built on an abstraction that lies even higher than Representation Theory and traditional Differential Geometry for historical reasons, on the one hand, and on the other, this formalism must be closely related to the most ubiquitous form of distribution in the physical sciences, the normal Gaussian density function.

This theoretical rendez-vous is located at the intersection between a certain form of Number Theory and the functional Gaussian formalism from which they emerge mainly as a system of tensors and operators. The most towering umbrella raised by this system of constructs is the principle of duality for all physical ontologies, which considers every object a reunion between an irreducible tangible tenet, or a *scalar* element, and an irreducible intangible tenet, or the *spatial void* element, both seized in a covariant relationship. This duality is captured in the following formula:

$$Q \times S = 1 \text{ or } Q = S^{-1} \text{ (4-1)}$$

where Q is the scalar and S the space tenet. The Quanto-Geometric function that incarnates the fullness of the principle of duality and codependency between the dual tenets is the following:

$$Q(s) = \pm \left| \frac{s^0}{\sigma\sqrt{2\pi}} \bullet e^{\left(\frac{-s^2}{2\sigma^2}\right)} \right|, s \neq 0 \text{ (4-2)}$$

where $Q(s)$ is the dependent variable portraying the scalar tenet and s the independent variable portraying the space tenet (Fig. 4.1).

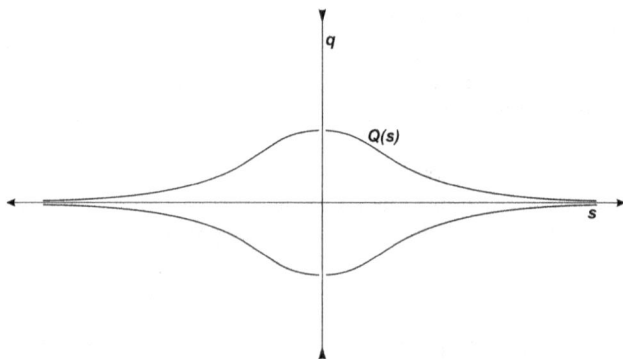

Fig. 4.1 Graph of the Quanto-Geometric function of one variable

Because in philosophy there cannot exist any absolute tangible scalar, nor any absolute intangible void, the function is expressed with a discontinuity at the q-axis introduced with the unconventional s^0 term. The q-axis is further mandated to surrogate pure scalar state. For the very same reason, the function displays an asymptotic behavior as s approaches 0, with the s-axis mandated to surrogate absolute pure void. Therefore our system of coordinates is not merely cardinal but it additionally and inherently reflects the background axiomatic proposition upheld in the Theory.

This set of directives and constraints result in the demotion or de-emphasis of the orthodox norms of distribution based on the standard deviation traditionally known to the Gaussian Function, as much as they call for a new form of normalization of the Function based on the conic or curvilinear symmetries delineated in its graphical pattern.

4.2. Natural Symmetry Groups

The association between the conic or curvilinear symmetries and their enclosing layered distribution Groups make for the most natural of symmetry groups, lying far above traditional transformation groups because they provide the normative basis for all forms of variance or transformations.

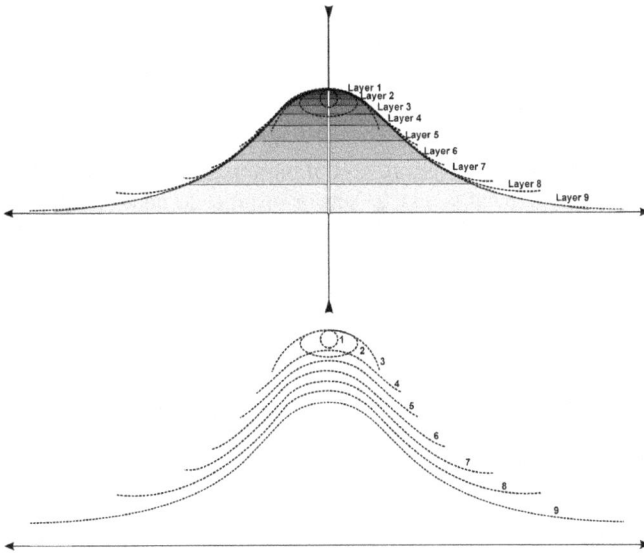

Fig. 4.2 The 9 essential layers of the Quanto-Geometric function with their associated manifolds or forms of symmetry

In this capacity, the Quanto-Geometric Function establishes a hierarchy of symmetry that ranks all known forms of symmetry in the following order (Fig.4.2):

1. circular form of symmetry

2. elliptic form of symmetry

3. parabolic form of symmetry

4. hyperbolic form of symmetry

5. hyperboloid forms of symmetry

The hyperboloid forms of symmetry regroup 5 distinct forms of symmetries that stand each in its very own right. The hierarchy so posits the existence of a total of 9 different fundamental forms of symmetry. This order of forms amounts indeed to an order of manifolds in the sense that the 4 conic forms and the 5 hyperboloid forms are the building blocks of all conceivable manifolds. In other words, the geometric composition of manifolds is strictly based on these unit-forms of symmetry, in the understanding that each one of them is fully embedded in the following one from unit-form 1 to unit-form 9.

4.3. Toward a Natural Hierarchy of Operators

Unit-norms are imprinted in the curvilinear metric or arc-metric of the Quanto-Geometric Function, giving rise to a set of 9 natural typified Quanto-Geometric tensors. These norms permeate the correlating sectionals or layers of the Function while giving rise to a set of 9 corresponding operators. To be sure, the referred norms constitute the binding agency that makes up the entire sets of numbers in all known forms of mathematical representation. The normal layers, as sets or groups, do no more than reflect the nature of the primal Numeric operators.

The Quanto-Geometric Operators represent ontological Norms applicable to physical observables in order to query their states and dynamics. This set of operators constitutes one of the higher orders of certainties that subsume all statistical order of uncertainties encompassing physical observables, that which is traditionally implemented from an integrative view of the function. We contrastingly give precedence to a layered view of the function thru its curvilinear sectionals.

4.4. Quanto-Geometric Coordinate System and its Metrics

In general, the metric tensor and its manifolds represent a broad program in differential geometry essentially aimed at describing the geometric properties of surface sheets. The seldom raised question of why curvature symmetries exist in 3 broad classes is one that finds natural fit within The Quanto-Geometric framework, these classes being:

1. Differentiable or Smooth continuous Manifolds (Riemannian)

2.   Finsler Manifolds

3.   Topologies (or Discrete Manifolds)

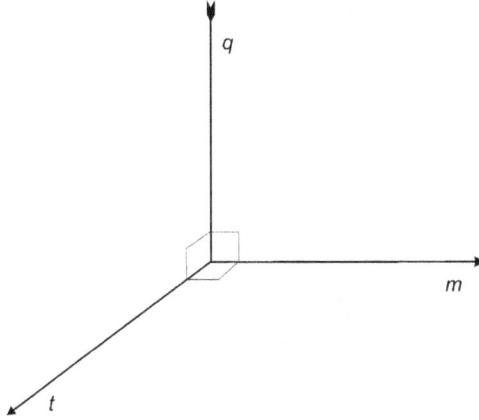

Fig. 4.3 Quanto-Geometric 3-space coordinate system

In Quanto-Geometry, the 3 dimensions of 3-space, despite the fact that they are isotropic to one another, portray 3 different metrics (Fig.4.3). In the 3-space background, two of the spatial degrees of freedom are associates or Abelian associates that are symmetric to one another. One represents the very archetype of the metric of symmetry and the other a kinematic metric. These two degrees of freedom, spatial depth *(t)* and spatial width *(m)*, together with their metrics are engenders of 3-space itself. The third degree of freedom, spatial height *(q)*, that is, is not a direct product of 3-space and its inner workings, but stems from interaction between innate 3-space and the Q_3 scalar (the value of $Q(s)$ at s_3 or 3-space). In representation, the third degree of freedom in 3-space becomes the scalar axis.

In 2-space background, space is two-dimensional, having a degree of freedom of height *(q)* and a degree of freedom of width *(s)*. The metric of one of the 2-space dimensions *(s)* represents a conflation of the distance (or symmetry) metric and the kinematic metric. This conflated metric is the innate expression of 2-space in and of itself. The second degree of freedom in 2-space, spatial height, that is, is not a direct product of 2-space and its inner workings, but stems from interaction between 2-space and the Q_3 scalar (the value of $Q(s)$ at s_2 in the graph). In representation, the second degree of freedom in 2-space becomes the scalar axis.

The Quanto-Geometric Theory advances both a 2-space Model or irreducible analytic image in 2 dimensions and a 3-space Model or irreducible analytic image in 3 dimensions, both of them being instrumental in the description of physical observables leading to the theoretical derivation of several different classes of fundamental physical constants. The important point here to bear in mind is that the nature of the coordinate system that one chooses to uphold in analysis is of utmost significance and that the familiar Cartesian system with the appended 4th dimension of time, for a 4-dimensional reference system, constitutes the source of many of the serious drawbacks and dead ends we have experienced in theoretical physics thus far. From the classical flat intuitive Cartesian system, the reference frame had acquired behavior in Einstein's theories of Relativity in very important ways. In the Quanto-Geometric framework, the dimensional background reference frame becomes intrinsically endowed with both ontological and phenomenological properties, which enhance its imaging capabilities as a system of representation and allow for very high towering observation and scrutiny of physical variables.

It is a fact of note that when importing mathematical constructs to physical observables, physicists have arguably not placed sufficient emphasis on the role of the very graphism of functions as expressions of the correlation between the independent and dependent variables at play. In our view the manifolds have become all too often abstract spaces obscuring interpretation instead of acting as direct elucidating tools. The geometric design of a function gives a direct measure of first-order correlation between its variables. Indeed much before the study of area under a curve by dividing it into infinitesimal boxes comes the study of curvilinear behavior of a curve by breaking it into infinitesimal sloped segments. The pertinence of this approach has led to the foundation of the entire mathematical science of differential geometry with important consequences in its physical application.

4.5. Specificities of the Quanto-Geometric Metrics

For all the importance and attention given to 3-space metric, 2-space Euclidean metric remains the most important of all metrics, if for no other reason because it constitutes the ultimate reference for all tensor categories arising from other metrics, such as the Ricci and scalar curvature tensors. Euclidean metric is in fact the most complex of all metrics being indeed the symmetry birthplace of all metrics.

All metrics, including Euclidean metric, are based on the distance metric which had been defined as stated below:

$d(x, y) = 0 \leftrightarrow x = y$, positing the identity of indiscernibles

$d(x, y) = d(y, x) \leftrightarrow x = y$, positing the property of symmetry

$d(x, z) \leq d(x, y) + d(y, z)$, positing sub-additivity or triangle inequality.

From the abstract axiomatic at its foundation, the Quanto-Geometric metrics are based on slightly different grounds in terms of their distance metric and on substantially different grounds when it comes to their differential metric. In Quanto-Geometry, the distance metric is based on the space between two points as stated below:

$$S : X \times Y \to \left] 0,+\infty \right[$$

The axiomatic so forbids the inception of 0 as a representation of pure void in physical reality on the claim that all ontologies are irreducibly dual in nature as formed by a scalar and a space element[3], all of which makes absolute void or 0 a natively vanishing entity. Furthermore, the nature of the distance element between two points is not only *geometrical*, as given in a straight line or a geodesic curve, but *dynamical* as well. The kinematic element is embedded in a degree of propagation coupled to the spread element incarnated in the distance figure. Therefore the *s*-metric predicates a *wavefunction metric* whose units are quantified in *wavenumbers*. Hence in Quanto-Geometry the distance between two points in space is a wavenumber. Going even further than conventional differential geometry, we pose that the Quanto-Geometric metric bears an etalon that is quantified in the standard deviation figure of σ, a distribution parameter, which makes it a seemingly running etalon.

4.6. Degrees of Freedom in Normalized Manifolds

The wavefunction dimension of the *s*-metric produces a scale of wavenumbers based on the σ etalon whose intervals are normalized and further represent unit-norms of distance and harmonic propagation. The primeval sequence s_n that produces this scale is:

$$\{s_n\} = \left\{ \frac{1-n}{10-n} \right\}, \quad \text{with } 1 \le n \le 9 \quad \text{(4-3)}$$

where n is a modulus of symmetry and/or degree of propagation. In its generalized application over the distance dimension of 2-space metric, its formulation becomes:

$$s = \pm\sigma\sqrt{\frac{(2n-2)}{10-n}}, \quad \text{(4-4)}$$

---

[3] We choose to apply the straightforward term *space* to the physical observable that is formally denoted *coordinate space* in the physical science literature by contrast to abstract spaces such as *momemtum space* or *phase space*. We do not peruse abstract fictitious spaces in Quanto-Geometry.

where s is the 2-space variable (See Section 5.5 for derivation of relation 4.4).

In that manner it yields the following super-imposed scale over the metric:

for $n = 1$, $s_1 = 0$

for $n = 2$, $s_2 = \pm\sigma/2$

for $n = 3$, $s_3 = \pm\sigma 2/7^{1/2}$

for $n = 4$, $s_4 = \pm\sigma$

for $n = 5$, $s_5 = \pm\sigma(8/5)^{1/2}$

for $n = 6$, $s_6 = \pm\sigma(5/2)^{1/2}$

for $n = 7$, $s_7 = \pm 2\sigma$

for $n = 8$, $s_8 = \pm\sigma 7^{1/2}$

for $n = 9$, $s_9 = \pm 4\sigma$

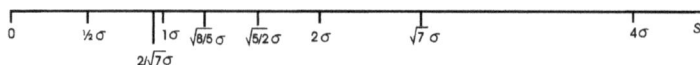

Fig. 4.4 Primeval sequence overscale of degrees of freedom laid on the s-axis

The s_i sequence of points represents the special wavenumber points that make up the scale (Fig.4.4). At the same time these points represent archetypical symmetry-breaking points delimiting the normal sectional curvatures of the curvilinear pattern of the Quanto-Geometric Function. They mark phase transitions from one form of the symmetry to another. If one cares to understand that all minimal sectional forms are typified forms of symmetry, then the sectional breakdown of the curvilinear design of the Quanto-Geometric Function ostensibly shows the natural hierarchy of symmetries ultimately dictated by 2-space metric. Those archetypical forms of symmetry constitute the basis of the sectional curvatures which grade the manifolds associated with all of the common differential geometry tensors. To make this point even clearer, we shall consider that degree of freedom Mod1 *(n = 1)* ultimately maps to the circular sectional of the curve thru interval]0, 1/2σ[of the *s*-metric. Below is reported the full suite of normal intervals on the metric with their correlation with the *n* degrees of freedom:

Interval 1 →]0, σ/2[, for *n = 1*

Interval 2 → [σ/2, 2σ/7$^{1/2}$[, for *n = 2*

Interval 3 → [2σ/7$^{1/2}$, σ[, for *n = 3*

Interval 4 → [σ, σ(8/5)$^{1/2}$[, for *n = 4*

Interval 5 → [σ(8/5)$^{1/2}$, σ(5/2)$^{1/2}$[, for *n = 5*

Interval 6 → [σ(5/2)$^{1/2}$, 2σ[, for *n = 6*

Interval 7 → [2σ, σ7$^{1/2}$[, for *n = 7*

Interval 8 → [σ7$^{1/2}$, 4σ[, for *n = 8*

Interval 9 → [4σ, ∞[, for *n = 9*

We have therefore established a direct relationship between the number of degrees of freedom and the number of focal points of the conic and hyperboloid forms making up the curvilinear design of the Function. This correlation gives us:

- One degree of freedom for one focal point of symmetry in the *circular* curvature

- Two degrees of freedom for two focal points of symmetry or foci in the *elliptic* curvature

- Three degrees of freedom for three focal points of symmetry or foci in the *para-bolic* curvature

- Four degrees of freedom for four focal points of symmetry or foci in the *hyper-bolic* curvature

- Five degrees of freedom for five focal points of symmetry or foci in the *entropic* curvature

- Six degrees of freedom for six focal points of symmetry or foci in the *enthalpic* curvature

- Seven degrees of freedom for seven focal points of symmetry or foci in the *paradoxical* curvature

- Eight degrees of freedom for eight focal points of symmetry or foci in the *perfect* curvature

- Nine degrees of freedom for nine focal points of symmetry or foci in the *infinite* curvature

There are good reasons to the epithets chosen for the hyperboloid forms of symmetry (5 to 9) which we address in the next Chapter. Suffice it to say for now that they stem from numeric signatures arising from the Number Stack that presides over the development of the framework. While the first 4 results may be of little surprise due to our familiarity with the conic sections, the degrees of freedom dictating the five hyperboloid forms of symmetry may likely come as a surprise due to their complete counter-intuitiveness. Very particularly, the last sectional, based on 9 degrees of freedom, practically portrays a linear element since this sectional infinitely lives in the neighborhood of the asymptotic *s*-axis. As such it represents, and counter-intuitively so, the most complex archetypical form of symmetry, whereas the circular sectional of curvature stands for the simplest archetypical form of symmetry, contrary to common wisdom in differential geometry. In that manner, this symmetry characterization makes explicit the reason that assists us in making the linear element with its stature of highest Modulus of freedom the most complex form of symmetry, as much as it explains why the linear element is indeed the irreducible in all metrics and why Euclidean-type frames are sufficient for the representation of the background of the Universal Arena.

4.7. Quanto-Geometric Primal Tensors

From the rectilinear distance metric we now move to the curvilinear metric that substantiates the curvilinear pattern of the Quanto-Geometric $Q(s)$ Function.

4.7.1 2-Space Tensors

The most irreducible representation of two points on the graph of the Quanto-Geometric Function implies a distance and a slope, which a vector entity perfectly incarnates. We may define the unit curvilinear metric by examining the behavior of a secant line thru any chosen point A on the curve (Fig.4.5).

The unit curvilinear metric is equivalent to the length or magnitude of the secant segment just prior to the secant line becoming a tangent to the curve at point A thru affine approximations of the secant to the tangential direction. We further clarify that in that scenario point B is the next dot in the curvilinear movement of the curve from point A and that segment [AB] is equivalent to the circular arc $A\widehat{B}$ (or $\hat{a}$) of *infinite* radius, an arc section which may be assimilated to a linear element for all intents and purposes. Therefore,

$$[AB]^2 = ds^2 + dq^2$$

$$[AB] = \sqrt{ds^2 + dq^2}$$

$$[AB] = \frac{1}{ds}\sqrt{ds^2 + dq^2} \cdot ds \text{, which becomes the arc-length:}$$

$$\hat{a} = \int_B^A \sqrt{\frac{ds^2}{ds^2} + \frac{dq^2}{ds^2}} \cdot ds$$

$$\hat{a} = \int_B^A \sqrt{1 + \frac{dq^2}{ds^2}} \cdot ds$$

$$\hat{a} = \int_B^A \sqrt{1 + \left(\frac{dq}{ds}\right)^2} \cdot ds$$

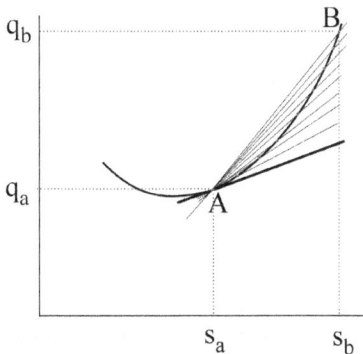

Fig. 4.5 Secant line approaching the tangential location
at one end point

We express the curvilinear metric or 2-space metric as:

$$g_i^2 = \int_B^A \sqrt{1 + \left(\frac{dQ(s)}{ds}\right)^2}\, ds \quad (5)$$

On the basis of this metric we may determine the length of any desired curvilinear sectional of the $Q(s)$ Function, which emerges as the measure of tensors or the tensile strength of the scalar-space correlation or covariance represented by the Function. The singlet tangent bundles that span each q-s normal interval, or the individual arc-lengths of these normals, give us the fundamental one-dimensional tensors that inhabit the curvilinear rundown of the Quanto-Geometric Function. We express the Quanto-Geometric 2-space tensors as integrals of the unit bundles, per expression (5), as follows:

Monolithic Tensor:

$$g_1^2 = \int_0^{\sigma/2} \sqrt{1 + \left(\frac{dQ(s)}{ds}\right)^2}\, ds$$

Parity Tensor:

$$g_2^2 = \int_{\sigma/2}^{2\sigma/\sqrt{7}} \sqrt{1 + \left(\frac{dQ(s)}{ds}\right)^2}\, ds$$

Associative Tensor:

$$g_3^2 = \int_{2\sigma/\sqrt{7}}^{\sigma} \sqrt{1 + \left(\frac{dQ(s)}{ds}\right)^2}\, ds$$

Power Tensor:

$$g_4^2 = \int_{\sigma}^{\sigma\sqrt{8/5}} \sqrt{1 + \left(\frac{dQ(s)}{ds}\right)^2}\, ds$$

Imaginary Tensor:

$$g_5^2 = \int_{\sigma\sqrt{8/5}}^{\sigma\sqrt{5/2}} \sqrt{1 + \left(\frac{dQ(s)}{ds}\right)^2}\, ds$$

Matrixial Tensor:

$$g_6^2 = \int_{\sigma\sqrt{5/2}}^{2\sigma} \sqrt{1 + \left(\frac{dQ(s)}{ds}\right)^2}\, ds$$

Rationalization Tensor:
$$g_7^2 = \int_{2\sigma b}^{\sigma\sqrt{7}} \sqrt{1+\left(\frac{dQ(s)}{ds}\right)^2}\, ds$$

Perfect Tensor:
$$g_8^2 = \int_{\sigma\sqrt{7}}^{4\sigma} \sqrt{1+\left(\frac{dQ(s)}{ds}\right)^2}\, ds$$

Transcendental Tensor:
$$g_9^2 = \int_{4\sigma}^{\infty} \sqrt{1+\left(\frac{dQ(s)}{ds}\right)^2}\, ds$$

Although the curve is a product of a 2-space reference frame or two-axis coordinate system, the curve itself is one-dimensional in the sense that the variations that make up the curvilinear pattern are realized only in the dimension of height of the coordinate system. Therefore the arc-lengths of the sectional curvature normals corresponding to the over-scale of the s-metric implanted in the s-axis constitute the preeminent quanto-geometric 2-space Tensors at the heart of the Theory.

4.7.2 3-Space Metric

When it comes to 3-space, per the two-variable expression of the Function *(expression 2-14)*, with its coordinate system bearing the properties that we have already discussed, the metric becomes bilinear, primarily informed by the two specialized spatial axes (Fig.4.6).

We can perfectly use the definition of the metric in 2-space as a root for the 3-space formulation of the metric. From 2-space to 3-space an axial transformation takes place, however, that splits the s-axis into two symmetric m- and t-axes, m and t being the new independent variables correlating to the one dependent variable Q(s). So therefore ds becomes *(dm, dt)*.

We now have the minimal segment of the curvilinear metric split into two equal segments which then form the triangular surface constituting the unit-surface for the metric. We may assimilate this unit-surface to the sectional curved surface of infinite radius (Fig.4.7). The area A enclosed in an equilateral triangle bears the following expression:

$$A = \frac{\sqrt{3}}{4}l^2 \text{ , } l \text{ being a side of the triangle.}$$

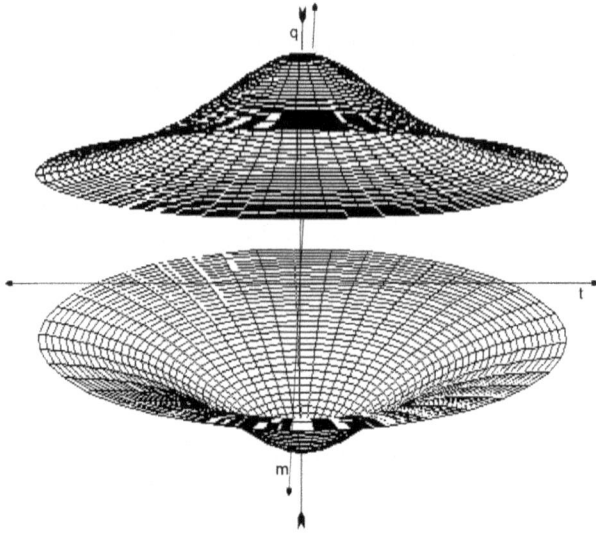

Fig. 4.6 A complete view of the graph of the Quanto-Geometric function of two variables *(m,t)*

Therefore the unit-surface of 3-space metric may be expressed as follows:

$$A = g_{im}^{3} = \frac{\sqrt{3}}{4} \left(\int_{a}^{b} \sqrt{1 + \left(\frac{\partial q}{\partial m} \right)^{2}} \right)^{2},$$

point *b* being the next dot to point *a* in the curvilinear movement of the generating curve on the partial *q-m* reference frame. Since *m* and *t* are symmetric, the 3-space metric may perfectly have the following alternate expression:

$$g_{it}^{3} = \frac{\sqrt{3}}{4} \left(\int_{a}^{b} \sqrt{1 + \left(\frac{\partial q}{\partial t} \right)^{2}} \right)^{2}$$

point *b* being the next dot to point *a* in the curvilinear movement of the generating curve on the partial *q-t* reference frame. The generating segments are duplet tangents that are the defining elements for the tangent bundles characterizing every point of the planar surface of 3-space manifolds, in accordance with the definition of 3-space manifolds in traditional differential geometry.

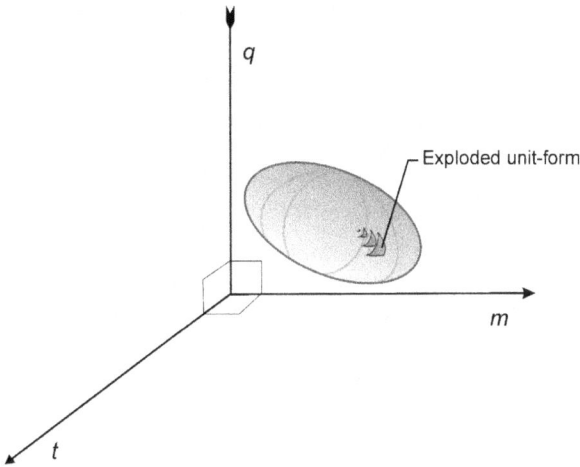

Fig. 4.7 Exploded view of a 3-space unit-form in an object

It is a notable fact however that the vast majority of parameters and the so-called fundamental constants of nature, which are at the foundation of 3-space or the 3-D universal domain, are artifacts of 2-space metric and its tensor derivates, 3-space metric contributing in a very limited manner in fashioning those quantities. One may intuitively deduce the expression of the 3-space tensors from that of the 2-space tensors, the list of which we will spare this exposé.

It is important to remember that despite the fact that we are codifying them as a length in following the method long in use in differential geometry, the basis metric units represent, beyond the mere length or arc-length element, a wave-length of a *wavefunction* (a wavenumber to be precise), as well as a *degree of propagation*. This follows from the properties initially incarnated within the Quanto-Geometric coordinate system. In that sense differential quanto-geometry represents a higher framework of generalization of which Riemannian geometry and non-Riemannian differential geometry are only particular cases.

At the height of its utility, the application of the transition map method in traditional differential geometry allows one to charter a course of interpretation of physical observables that is as comprehensive and complete as afforded by the atlas of representation. However, we do not apply this technique in Quanto-Geometry because the mathematical-physics nature of our coordinate systems for every *n*-space, with the properties expressly subsumed in their axes, affords us to directly compute the quantities borne out by the states and dynamics of the physical observables, or alternatively by application of the Quanto-Geometric Operators. All of which traditional Differential Geometry is unable to materialize. One can already see that Quanto-Geometry solidly encapsulates both Euclidean Geometry and Differential Geometry.

4.8. Quintessential Tri-Valent Tensor

In continuing to scrutinize it along the line of differential geometry formalism, one more important characterization springs from the geometric design of the Function. First order derivation, which stands for the second order of the quanto-geometric correlation between the scalar and space tenets (the function itself being the first correlational order), gives us three broad sectors dividing the graphical pattern:

In one phase → $\dfrac{dq}{ds} > 1$ or in 3-space $\left.\dfrac{\partial q}{\partial t}\right|_{m} > 1$ and $\left.\dfrac{\partial q}{\partial m}\right|_{t} > 1$, which translates into the scalar tenet dominating or eclipsing the space tenet in the correlation.

In the second phase → $\dfrac{dq}{ds} < 1$ or in 3-space $\left.\dfrac{\partial q}{\partial t}\right|_{m} < 1$ and $\left.\dfrac{\partial q}{\partial m}\right|_{t} < 1$, which translates into the space tenet dominating or eclipsing the scalar tenet in the correlation.

In the third or mid-phase → $\dfrac{dq}{ds} = 1$ or in 3-space $\left.\dfrac{\partial q}{\partial t}\right|_{m} = 1$ and $\left.\dfrac{\partial q}{\partial m}\right|_{t} = 1$, which means equated covariant influence between the scalar tenet and the space tenet. The first-order derivative monotonically increases in phase 1, monotonically decreases in phase 2 and remains invariant in the mid-phase.

In the first sector the scalar overwhelms its space partner in the covariant relationship. In the 3rd sector, the space partner overwhelms the scalar. And in the 2nd or mid-sector the covariant relationship between the two partners is equalized. The nature of these differential sectors is equivalent to *singlet* tangent bundles. We denote them as a whole the *Quintessential Trivalent Tensor* in the understanding that the latter represents 3 different tangent bundles or 3 component trivalent tensors, namely the *Quantum-dominant* or *Scalar-Dominant* Tensor, the *Space-dominant* Tensor and the *Quadratic* Tensor. This characterization is one of the most pivotal concepts of the Quanto-Geometric framework with far reaching consequences.

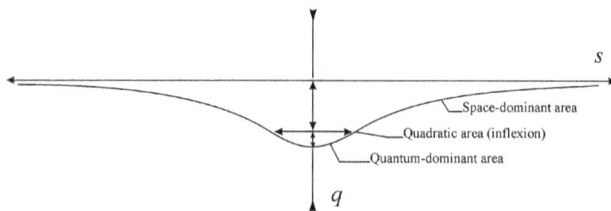

Fig. 4.8 The three sectors of the Quanto-Geometric Function that grandfathers the Tri-Valent Operator

Hyper Quantum Quintessence is a generator portending to the formation of closed spaces. Hyper Space Quintessence is a generator of surface sheets or open spaces. Hyper Quadratic Quintessence is a generator of semi-closed/semi-open spaces. If the first generator has its way, it will make everything circular in creating volumes, whether 2-dimensional (circle), or 3-dimensional (sphere) thru the nth dimensional order. If the secondly mentioned generator has its way, it will make everything straightly linear by creating flatness, whether 2-dimensional (planes) or plane-like in all dimensional orders. And if the thirdly mentioned generator has its way, it will turn everything into thickness in creating layers by giving equalized weight to vertical and horizontal deploys.

The determinations of this symmetric order are of paramount significance. They are not quite in line with common wisdom in traditional differential geometry. For instance, one may argue that a circle is a plane figure. However, if we define a volume a closed space or a closed-like space, per our familiarity with conformal 3-space, in a two-dimensional world the circle must be homeomorphic to the sphere as the simplest form of closed space or volume. While the charts and transition maps are probably a useful tool to study the parallels between 2-space (or R^2) manifolds and other higher n-space (or R^n) manifolds, they do not establish the higher order of symmetry at the foundation of the distribution of manifolds. Tri-Valent Quintessence is an ultra-structural categorization that reveals the higher class of symmetry to which manifolds and tensors belong, in a manner that is similar to group covers in Group Representation Theory. In that sense Tri-Valent Quintessence reveals the three super-classes that regroup manifolds and tensors in differential geometry. It is by virtue of Tri-Valent Quintessence that three and only three broad classes or super classes of these elements are in existence.

In establishing a parallel between our framework and Category Theory abstractions, we may point out that there is a homeomorphic relationship between the order of surface sheets in 2-space and the class of Riemannian (differentiable) manifolds in 3-space. There is a homeomorphic relationship between the order of closed spaces in 2-space and the order of topological manifolds, generator of volumes, in 3-space. There is a homeomorphic relationship between the order of semi-closed/semi-open spaces or layers and the class of Finsler manifolds in 3-space. Tri-Valent Quintessence starts in 2-space and propagates thru n-space by homeomorphism, keeping in mind that a specific axis transformation is required to transition from one space form to another. This homeomorphism is a result of the fact that at each subsequent n^{th}-space the spatial entity develops a richer multiplet state than the previous.

4.9. Quanto-Geometric Primal Operators

In a nutshell, consider this: arithmetic calculus is based on 4 simple operators of primarily passive agency:

- The Monolithic Operator that promotes the method of addition [+].

- The Parity Operator that promotes the method of Sci-Parity or Se-paration, which we call subtraction [-]. The method of division [/] for its own is a poster child to the method of subtraction.

- The Associative Operator which promotes the relational method of multiplication [×].

- The Radical Operator which promotes the methods of square root (or roots) [√] and squaring [x]^2.

There exist 5 additional Operators, more complex and primarily of active agency, that are at the root of the calculational methods put forth in functional calculus otherwise called functional analysis. They are:

- The Complex Operator which promotes calculus of the imaginary.

- The Harmonic Operator which promotes matrix calculus.

- The Rationalization Operator which promotes irrational calculus.

- The Golden Operator which promotes algebraic calculus.

- The Transcendental Operator which promotes serial calculus.

The norms that are rooted in the Quanto-Geometric Number Stack (Chapter 3) create not only the class of Normalized Tensors that we have previously studied but the parallel class of Normalized Operators above enumerated, this thru the Quanto-Geometric Function. One can see that the natural bounds of the span of *operation* are the punctual Monolith at one end of the spectrum and the Infinite series at the other end. The Quanto-Geometric formal expression for these irreducible put under the same umbrella not only the functionally-woven Operators (the last 5) but the arithmetic Operators as well.

From mathematical study undertaken over the Quanto-Geometric Function (Chapter 2), the ensuing generic expression of the Quanto-Geometric Operators is expressed below:

$$\|\Psi\| = \left[\frac{-Q^2}{2q} \right] \bullet \left[\frac{2}{ae^{\left(\frac{2a+b^2}{4}\right)} + be^{\left(\frac{2a+b^2}{4}\right)} + ce^{\left(\frac{2a+b^2}{4}\right)}} \right]$$

where:

q is the scalar and equates $e^{a/2}$ or: $q = e^{\left(\frac{a}{2}\right)}$

Q is the primal quanto-geometric momentum $q \times s$ or: $Q = q \times s$

a is a coefficient related to q and equates $2 \ln q$ or: $a = 2 \ln q$

b is a coefficient related to the standard deviation parameter and equates $\ln \sigma$ or: $b = \ln \sigma$

c is the constant $\ln 2\pi$ or: $c = \ln 2\pi$

The idea that originally motivated the inception of functional Operators in mathematical science was the question of finding the shape of a vibrating string in a musical string instrument for a certain family of notes. The string is fixed at its boundaries and has a certain length. Those are the invariants that condition the variance that will be given in the ensuing standing waves the chord will be subject to. It is by posing the problem in terms of a differential operator denoting the displacements of the infinitesimal portions of the string that we obtain a set of eigenfunctions and eigenvalues to this operator as the solution to the problem. The operator is of the form of a second-order derivative:

$$\frac{\partial^2}{\partial x^2}$$ in the length dimension

and

$$\frac{\partial^2}{\partial t^2}$$ in the time dimension

acting on a function h of the displacement and time variables: $h(x,t)$.

Likewise we may pose the question of finding the set of wavefunctions or harmonic oscillations experienced by a space spread (field) in codependent relationship with a scalar partner as the most generic question surrounding the ontology of any physical object. To that aim we may bring the Quanto-Geometric Operator to the fore, recognizing that the family of functions embedded in the Quanto-Geometric Function are the eigenfunctions along with their eigenvalues that are proper to the generic Quanto-Geometric Operator.

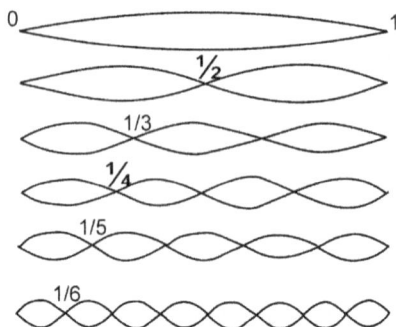

Fig. 4.9 Overtone waveforms of a vibrating string or drum as eigenfunctions and eigenvalues

The harmonic pattern in Fig. 4-10 is very similar to the spectral pattern (Fig. 4-9) imputing the analytic notion of normed function space with intervening Operator, except for a few important distinctions. The distribution in Fig.4.10 portrays a sequence of harmonic waveforms involving different families of functions. Every eigenvalue is thus associated with a waveform of a distinct nature from the next. The eigenvalues are not just a coefficient to the period of the waveform or even to the function but come as cosets implicating both the wavenumber (or wavelength) and the scalar density associated with the waveform. What this characterization reflects is that the Quanto-Geometric eigenspectrum represents a level of generalization that lies far above abstract Hilbertian function space, that which plays a central role in traditional Operator analysis.

Every layer of waveforms represents in fact a whole functional domain, which then might host different harmonic spectra of Hilbert spaces. Every functional layer of the Quanto-Geometric spectrum constitutes the Primitive or Generator of a normed function space, or, if you will, an Operator primitive. One of the properties that significantly distinguish the Quanto-Geometric Operator from Hermitian Operators is the ontological character subsumed in the variables of their function basis. The main quality of Hermitian Operators is the geometric orthogonality of their L^2 eigenfunction basis. Comparatively, the orthogonality of the Quanto-Geometric Operators rests on the dual ontological set inscribed in the unit function of their function basis, while their norm is given by the typology of the set. Expression *(4-1)*, which mandates the scalar tenet of the function and its space tenet to be the reciprocal of one another, is the most iconic representation of this orthogonality. In that sense there is homomorphism between the orthonormal criteria advocated in both frameworks.

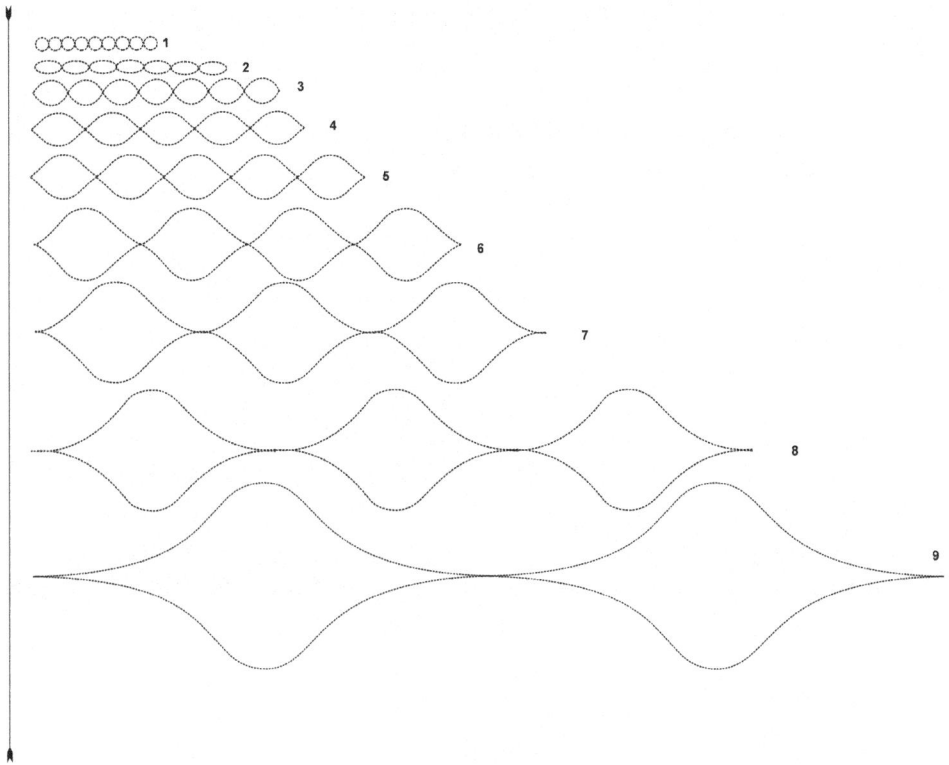

Fig. 4.10 Waveform spectral of families of functions embedded in the Quanto-Geometric Function

Hilbert orthogonality is between an element f and its conjugate g or $\bar{f}$ forming an inner product. The duality is between two conjugates and their relationship is measured in terms of a geometric angular element, specifically the cosine of the angle formed by their representative vectors. The angle must be 90° or the cosine of *null* value for orthogonality to hold in the vector space. In Quanto-Geometry, the function space is an entire functional domain and every eigenfunction primitive of the different domains inherits the qualities set forth by the towering Quanto-Geometric relation expressed in *(1)*, which in and of itself is seminally orthonormal.

In summary, the Quanto-Geometric functional system is based on a 9-dimensional real-coordinate space, with each of the 9 eigenfunctions in the role of generator of a family of functions with typified *eigenforms* (circle, ellipse, etc.). The generic Quanto-Geometric Operator is the master key to the 9-dimensional eigenspectrum and grandfathers 9 typified Operators, each acting on or adjoined to the specific functional family in their domain. All known or theorized abstract spaces are lower-level entities of partial or limited scope which pertains to the different functional domains germane to the Quanto-Geometric eigenspectrum.

4.10. Expressions of the Nine Specific Quanto-Geometric Operators

Consonant with the above, we may now express the generic Quanto-Geometric Operator as follows:

$$\left\| \Psi \right\|_{a_i b_i} = \left[\frac{-Q^2}{2q} \right] \bullet \left[\frac{2}{a_i e^{\left(\frac{2a_i + b_i^2}{4} \right)} + b_i e^{\left(\frac{2a_i + b_i^2}{4} \right)} + c e^{\left(\frac{2a_i + b_i^2}{4} \right)}} \right]$$

This Operator applies to the whole eigenspectrum portrayed in Fig.4.10. Every sub-index i in $a_i b_i$ represents a q-s set, yielding the following suite of generators or primitive Operators:

Monolithic Operator:

$$\left\| \Psi \right\|_{a_1 b_1} = \left[\frac{-Q^2}{2q} \right] \bullet \left[\frac{2}{a_i e^{\left(\frac{2a_1 + b_1^2}{4} \right)} + b_i e^{\left(\frac{2a_1 + b_1^2}{4} \right)} + c e^{\left(\frac{2a_1 + b_1^2}{4} \right)}} \right]$$

Parity Operator:

$$\left\| \Psi \right\|_{a_2 b_2} = \left[\frac{-Q^2}{2q} \right] \bullet \left[\frac{2}{a_2 e^{\left(\frac{2a_2 + b_2^2}{4} \right)} + b_2 e^{\left(\frac{2a_2 + b_2^2}{4} \right)} + c e^{\left(\frac{2a_2 + b_2^2}{4} \right)}} \right]$$

Associative Operator:

$$\left\| \Psi \right\|_{a_3 b_3} = \left[\frac{-Q^2}{2q} \right] \bullet \left[\frac{2}{a_3 e^{\left(\frac{2a_3 + b_3^2}{4} \right)} + b_3 e^{\left(\frac{2a_3 + b_3^2}{4} \right)} + c e^{\left(\frac{2a_3 + b_3^2}{4} \right)}} \right]$$

Radical Operator:

$$\left\|\Psi\right\|_{a_4 b_4} = \left[\frac{-Q^2}{2q}\right] \bullet \left[\frac{2}{a_4 e^{\left(\frac{2a_4 + b_4^2}{4}\right)} + b_4 e^{\left(\frac{2a_4 + b_4^2}{4}\right)} + ce^{\left(\frac{2a_4 + b_4^2}{4}\right)}}\right]$$

Imaginary Operator:

$$\left\|\Psi\right\|_{a_5 b_5} = \left[\frac{-Q^2}{2q}\right] \bullet \left[\frac{2}{a_5 e^{\left(\frac{2a_5 + b_5^2}{4}\right)} + b_5 e^{\left(\frac{2a_5 + b_5^2}{4}\right)} + ce^{\left(\frac{2a_5 + b_5^2}{4}\right)}}\right]$$

Matrixial Operator:

$$\left\|\Psi\right\|_{a_6 b_6} = \left[\frac{-Q^2}{2q}\right] \bullet \left[\frac{2}{a_6 e^{\left(\frac{2a_6 + b_6^2}{4}\right)} + b_6 e^{\left(\frac{2a_6 + b_6^2}{4}\right)} + ce^{\left(\frac{2a_6 + b_6^2}{4}\right)}}\right]$$

Rationalization Operator:

$$\left\|\Psi\right\|_{a_7 b_7} = \left[\frac{-Q^2}{2q}\right] \bullet \left[\frac{2}{a_7 e^{\left(\frac{2a_7 + b_7^2}{4}\right)} + b_7 e^{\left(\frac{2a_7 + b_7^2}{4}\right)} + ce^{\left(\frac{2a_7 + b_7^2}{4}\right)}}\right]$$

Golden Operator:

$$\left\|\Psi\right\|_{a_8 b_8} = \left[\frac{-Q^2}{2q}\right] \bullet \left[\frac{2}{a_8 e^{\left(\frac{2a_8 + b_8^2}{4}\right)} + b_8 e^{\left(\frac{2a_8 + b_8^2}{4}\right)} + ce^{\left(\frac{2a_8 + b_8^2}{4}\right)}}\right]$$

Transcendental Operator:

$$\|\Psi\|_{a_9 b_9} = \left[\frac{-Q^2}{2q}\right] \bullet \left[\frac{2}{a_9 e^{\left(\frac{2a_9+b_9{}^2}{4}\right)} + b_9 e^{\left(\frac{2a_9+b_9{}^2}{4}\right)} + ce^{\left(\frac{2a_9+b_9{}^2}{4}\right)}}\right]$$

Let me note in passing that the term Golden for type-8 Operator is not a phenomenological attribute but a tribute to the key role played by the Golden Ratio in fomenting algebraic calculus, per numeric analysis of inner structures of the Higher Number Stack exposed in Chapter 3.

The functional domain associated with the Monolithic Operator is the family of circles or circular functions.

The functional domain associated with the Parity Operator is the family of elliptic functions.

The functional domain associated with the Associative Operator is the family of parabolic functions.

The functional domain associated with the Radical Operator is the family of hyperbolae.

The functional domain associated with the Hyperboloid Operators consists in the five distinct families of hyperbolic functions.

In support of this view, a direct exponential expression of each of the first 4 families of functions, the so-called conic functions, is well known in different forms. The root Jacobi theta function that gives rise to the family of so-called *elliptic functions* has the following exponential expression:

$$\theta_{ab}(x\,|\,\tau) = \sum_{n \in Z} \exp\left[\pi i\,\tau\left(n+\frac{a}{2}\right)^2 + 2\pi i\left(n+\frac{a}{2}\right)\left(x+\frac{b}{2}\right)\right] \quad \text{where } \tau \text{ is the modulus}$$

or excentricity and $a,b = 0,1$.

The general class of Jacobi elliptic functions may be viewed as the root for the exponential expression of all 4 known conic. Effectively those functions are closely related to the geometry of *elliptic curves*, in particular the familiar conic ellipse which historically motivated their development amidst the quest for methods to compute the arc-length of the latter. Further their rectangular arguments can be parameterized to produce 4th-, 3rd- and particularly 2nd-degree polynomials, which are conic parabolic, while they may drop down to circles as well. Needless to point out their relationship with the hyperbola, since they hold a parallel trigonometric expression.

It is noteworthy that these exponential expressions almost always include a complex element, while they may alternatively include matrix terms as well. In that regard, the quanto-geometric spectrum duly shows that there is no direct path from exponential calculus to quadratic calculus on down. Thus the expression of a specific level of the stack from elements of a higher level stratum must include any intermediate calculational figure.

4.11. Self-Adjoint Quanto-Geometric Operation

We must stress that the functional representation of these Operators does not supersede or do away with the arithmetic operators but simply re-formulate them. The first 4 expressions of these Operators, namely the Monolithic Operator, the Parity Operator, the Associative Operator and the Radical Operator are respectively completely equivalent to the familiar arithmetic operators. Consider this: when we see for instance a set of 2 chairs in a corner of a room and another set of 3 chairs in the next corner of the room, and say there are 5 chairs in the room, which we express as 2 [+] 3 = 5, what we are doing as observers or analysts is applying the Monolithic Norm to the scalar state of these observables by which we aim to conceptualize them as **one (mono)** group of objects. Note that there is operation as well on the spatial state of these objects from the spatial monocentric symmetry element encapsulated by the Operator, which we normally do not consider part of the summation operator in elementary Number Theory, and perhaps wrongly so. If we further physically join these chairs, then we have realized or made explicit the *transformation* implicit in the Monolithic Norm. Therefore it is perfectly correct to write:

$$[+] \equiv \left\| \Psi \right\|_{a_1 b_1}$$

to mean equivalency between the arithmetic operator of summation and the Monolithic Quanto-Geometric Operator. Likewise we may apply the same rationale to the phenomenology of these objects in the evoked context insofar as the two subsequent arithmetic operators, which would give us:

$$[-] \equiv \left\| \Psi \right\|_{a_2 b_2}$$

$$[\times] \equiv \left\| \Psi \right\|_{a_3 b_3}$$

If there were only 3 chairs in one and only one corner of the room, under the condition that there are 9 people needing accommodation in the room, we might consider as agents a replicate of these 3 chairs in two other desired corners of the room. In thinking of the original group of chairs, we might conceptualize a squaring of the scalar state of those chairs by applying one of the conjugate components of the Radical Operator to their scalar state, specifically the Square, for the desired transformation of obtaining more of these chairs. Thus we may justifiably write:

$$\left[\sqrt[n]{} : \left\langle m^n \right\rangle \right] \equiv \left\| \Psi \right\|_{a_4 b_4}$$

to mean equivalency between the arithmetic operator of squaring and the Radical Quanto-Geometric Operator. The reason why the 5 hyperboloid Operators do not have simple arithmetic replicates is because of the preeminent role played by the exponential wavefunction component in the transformations realized by these operators, which is expressed in the wealth of methods associated with their realization. The great diversity of these methods constitutes the province of Functional Analysis. However, when a molecular entity of the biosphere such as a gene is able to infinitely and exactly replicate itself across an endless number of generations, one can be sure that a Transcendental Operator is at work within the seminal structure of this object, whereby the wavefunction plays an absolutely preeminent role vis-à-vis that one played by the scalar tenet of the Operator.

In mathematical science, we have not thus far considered the many different functional methods in Functional Analysis as calculational operators deserving nomenclature based on operator symbols similar to the ones adopted for the four arithmetic operators. The sum of these methods pertain to the hierarchy of the hyperboloid Operators. And therefore all nine Quanto-Geometric Operators, in their condition as direct expression of arithmetic and calculational operators, are all primal *self-adjoint* operators.

4.12. Quanto-Geometric Linear and Non-Linear Operators

The Transcendental Operator presents a special attribute that uniquely distinguishes it from all others, in particular from the other functional Operators of the spectrum. With its inherent methods of derivative and integrative operation, this entity presents the unique characteristic of linearity primarily due to the exponential nature of its quanto-geometric eigenfunction. We attribute the character of linearity to an operator A acting on a function f whenever the operator returns from the transformation the same exact function except for multiplication by a scalar constant, as expressed below:

$$Df = af$$

where D is the Operator, f the function and a the scalar constant. Because the exponential function is the only known one that is at the same time its own derivative and its own integral, it becomes the eigenfunction of the derivative operator and the integral operator.

Therefore the function space originated in the transcendental layer of the Quanto-Geometric Function, being the province of exponential calculus, becomes the haven of Linear Operation. Analysis of the structure of this layer of the Quanto-Geometric Function is quick to reveal that because of the eternal vicinity of the curve to the asymptotic s-axis in that sector, the Function behaves linearly or with effective linear symmetry therein, due to straight parallelism with the asymptote. This essential character of Linear symmetry reflects on the function space and thus onto the Operator, making it not an abstract but a real character.

All other Quanto-Geometric Operators (types 8 to 1) are non-linear because the symmetry *mani-folded* in their function spaces is curvilinear, since those spaces are strictly made up of conic and more or less curved hyperboloid sectionals.

4.13. Inner Ontological Properties of the Quanto-Geometric Operators

The very specific values taken by the set *[a, b]* in the expression of these Operators determines the Primitive identity or Norm of the operator, in other words one among the 9 Quanto-Geometric *normals*. It cannot be overstressed that none of these function spaces is an abstraction or an abstract Hilbert space, but represents each a property of real coordinate space. By the same token, and in catering to the smooth span of these 9 basis function spaces, one can inarguably concord with the alternative designation of Grand Eigenfunction that we have attributed to the Quanto-Geometric Function itself.

Delving somewhat further into the structure of the generic Quanto-Geometric Operator, let us note that the leftmost term in its expression may be assimilated to an energy density term that establishes the order of <u>potential</u> energy in an object as its highest scalar state, a static state:

$$\left[-\frac{Q^2}{2q} \right].$$

The next term at the left side is the wavefunction term that grandfathers the kinetic energy of the object in its most elevated form while defining its form of symmetry:

$$\left[\frac{2}{a_i e^{\left(\frac{2a_i+b_i^2}{4}\right)} + b_i e^{\left(\frac{2a_i+b_i^2}{4}\right)} + ce^{\left(\frac{2a_i+b_i^2}{4}\right)}} \right].$$

One can readily notice the hyperbolic form of expression in that term as the ratio between 2 and a sum of exponentials typical of the hyperbolic functions, very particularly the *sech* function. Each of the two terms gives us a composition between q and s, but with one of the two variables with a preeminent weight in each term, thereby giving the specific term its final ontological identity. In that regard, we must remember that e is the numeric expression of space in transcendentalization and that σ is a value of s.

4.14. A Parallel Between the Generic Quanto-Geometric Operator and the Hamiltonian Operator

When scrutinizing the nature of the generic Quanto-Geometric Operator, it is interesting to compare it to the Quantum Mechanics Hamiltonian:

$$\left[-\frac{\hbar^2}{2m} \cdot \frac{d^2}{dx^2} + V \right] = H \quad (7)$$

The core term that conforms the *operation* in the Hamiltonian is:

$$\left[-\frac{\hbar^2}{2m} \cdot \frac{d^2}{dx^2} \right],$$

the kinetic energy Operator, which equation *(7)* above drops down to when the potential energy V equals *0*. The Hamiltonian Operator is indeed a reflection of the Quanto-Geometric operator canon in that it is a composition between an energy density term, the first one at left of the expression, identical in form to the Quanto-Geometric scalar state term, and a *wavefunction* term, the second one, here expressed as the second-order derivative of a *function*. Rewriting the core term as expressed below makes even more apparent the parenthood relationship between the Hamiltonian and the generic Quanto-Geometric Operator:

$$\left[-\frac{\hbar^2}{2m}\right] \bullet \left[\frac{d^2}{dx^2}\right].$$

It is by virtue of narrow compliance to the higher canon that the eigenfunction of the Hamiltonian Operator turns out to be:

$$\phi_m(\varphi) = \frac{1}{\sqrt{2\pi}} \cdot e^{im\phi},$$

a functional expression in polar coordinate that is a faithful image of the Quanto-Geometric Function. So it is not just because the Hamiltonian is a linear operator. In particular, the wavefunction term of the generic Quanto-Geometric Operator contributes to a significant extent to clarify the nature of the Quantum Mechanics wavefunction, a subject of historically unrelenting controversy. The wavefunction term of the Quanto-Geometric Operator initially instantiates the construct that the Quantum wavefunction is not an abstract vector space, but a real-valued *eigenform*, specifically one of 9 possible distinct universal *eigenforms*. Therefore, at the very least it proposes that:

1. A wave of any nature is not necessarily, if at all, an expression in *time* of the eternal sinusoid but a specific spatial *eigenform* of which the sinusoid as a circular manifold is only a class.

2. A wave is dynamically a complex function of two quantities: one that is both coupled to and expressed by the *eigenform* with vibrational or oscillatory manifestation, and a quite separate entity which is its degree of propagation, *0* being a possible value of the latter.

3. The very fabric of the void is the originator of all harmonic dynamics thru the modality of *de-ploys* of its intangible continuum. In that regard, consider that a differential operator of any n^{th} order, progeny of the spatial term of the Quanto-Geometric Operator, is an entity of active agency engendered thru auto-dynamism, as it were, of a metaphoric secant line to a curve moving to its single-point tangential location.

These notions are all implicit in the Quanto-Geometric *wavenumbers* in their *s*-space form or *(m,t)*-space form, the independent variable to the Quanto-Geometric coordinate system respectively in 2-space and 3-space. There is for certain far more to the notion of wavefunction than we have expounded here. Much more in Chapter 7.

With the analytical order of Operators concluding with the notion of *eigenforms*, we have circled back to the analytic order of Tensors and Manifolds, in a tight demonstration of the consistency of the whole Quanto-Geometric system of constructs. Effectively there is within the framework a direct Operator expression of the normalized Tensors. However we will spare it to this exposé.

4.15. Quintessential Tri-Valent Operators

Similar to the Tensor order, the triplet partition of the Quanto-Geometric Function equally has an echo in the order of *operation* subjacent to the Function. Despite the fact that we have repeatedly used the functional figure of first-order derivative to extol the quintessential triplet characterization, it is important to understand that the latter is not subject to nor is it intrinsically defined by functional derivatives.

It remains obvious that the first 3 conic families of functions that are ostensibly manifest in the graphical rundown of the Function, namely the circle family, the elliptic family and the parabolic family, have in common a "closed" form of symmetry, explicit in the orientation of the tangential angle to that curvilinear sector of the Function, while the last 5 families share an "open" form of symmetry manifest in the orientation of the sweeping tangential angle as well. We say that each normal sectional is a family because they are all parameterized on the basis of a, b and σ, and indeed beyond those parameters, as previously explained, to produce their own spectrum. The quadratic sector is a Hilbert space as well for the same reason. The partial function space represented by each sector is smooth and orthonormal, and therefore represents each the partial spectrum of eigenfunctions of an adjoint operator.

We denote these Operators:

1. The Hyper Quantum Operator for the first sector

2. The Hyper Space Operator for the second sector

3. The Hyper Quadratic Operator for the mid- sector

The intervention of each of these three operators on physical observables respectively causes an **accrued** scalar densification, an **accrued** linearization or *superficialization* of the object and an **accrued** form of ambivalence or mirror imaging within the same.

While we normally and principally use functional linear operators, of Hermitian nature or else, to query state and dynamics of observables, the entire suite of Quanto-Geometric Operators are equally applicable either to *query* the unknowns about observables or to *inform* an observable with a quality emanating from the Norm of the Operator, thereby *creating* a *transformation*. In the latter condition, these functional Operators exert their *transformative norm* in a way that is relatively similar to the arithmetic operators.

4.16. Universality of the Quanto-Geometric Operators

It is important to realize that the Quanto-Geometric Operators apply not only at the level of the quantum realm but at all subsequent levels of the development of physical matter.

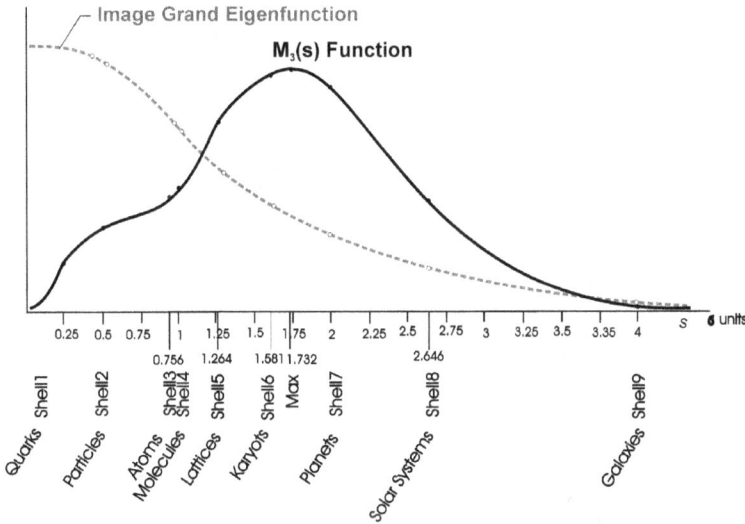

Fig. 4.11 An incipient view of the suite of Shells in the transverse order

In their genericity, they represent an essential scheme that informs all ontologies thru the course of the buildout of the several different Shells of matter as they appear before us, namely the Quark Shell, the Particle Shell, the Atom Shell, the Molecular Shell, the Material or Lattice Shell, all the way up to the Galactic Shells.

Therefore, on that account, they duly inform the ontology of objects in the Astral Shell (individual planets, moons or satellites, comets and stars), habitat of all Karyots. Classical physics has ascertained the following notable quantities from dynamical observables in our immediate environment:

Linear Kinetic Energy

$$From\ K_l = m\ v^2\ /2,\ p = m \times v \rightarrow\quad K_l = p^2\ /2m,$$

where p is the linear momentum of the object in motion.

Rotational Kinetic Energy

$$\text{From } K_r = I \, \omega^2 /2, \quad H = I \times \omega \rightarrow K_r = H^2 /2I,$$

where H is namely the rotational momentum of the object in motion.

Energy in Magnetic Flux of Inductor

$$U = L \, i^2 /2, \quad Li = N\Phi \rightarrow \quad U = (N\Phi)^2 /2L,$$

where the quantity $N\Phi$ is defined as the flux linkage of the inductor.

Energy Density in an Inductor

$$U = \mu_0 n^2 \, i^2 /2, \quad \beta = \mu_0 n \, i \rightarrow \quad U = \beta^2 / 2\mu_0$$

where β is the strength of the magnetic field set up by a current-carrying inductor.

Energy Stored in a Capacitor

$$U = C \, V^2 /2, \quad Q = CV \quad \rightarrow \quad U = Q^2 /2C,$$

where Q is the total amount of charge stored on the plates of a capacitor.

Energy Density in Field of a Charged Surface

$$U = \varepsilon_0 \, E^2 /2, \quad \sigma = \varepsilon_0 \, E \quad \rightarrow \quad U = \sigma^2 /2\varepsilon_0,$$

where σ is defined as the charge density on a surface.

Kinetic Energy in Liquid Flow

$$K = \sigma v^2 /2, \quad F = \sigma v \quad \rightarrow \quad K = F^2 /2\sigma,$$

where F is the flow rate of a liquid.

Elastic Energy Stored in a Spring

$$U = k \, x^2 /2, \quad F = k \, x, \quad \rightarrow \quad U = F^2 /2k$$

The second expression is known as Hooke's law, and F the restoring force of a spring.

Save the negative sign, these quantities are completely conformal to the scalar component of the generic Quanto-Geometric Operator. It follows from this treatment that the Newton's First Law approach, which essentially creates the momentum-like figure as a product between the scalar and its motion, is only phenomenological in appearance (or descriptive of exogenous dynamics otherwise called interactions) in some cases. From the Quanto-Geometric vision the momentum figure emerges instead as an ontological quantity about the internal composition of an object as part of its scalar component. The Quanto-Geometric Operator predicates that what appears to be the compactified component in an observable residing within any Shell of matter, including our own Shell of habitation, that which we call mass or quantum or energy packet is never a pure monolith but always a density quantity in the form of an integral sum or packet modeled by $\left[-\dfrac{Q^2}{2q} \right]$. It

represents the first ontological level of the scalarity of the observable. In that regard, the energy nomenclature for the above reported quantities would be quite misplaced if we do not consider the notion of energy as the very primal nature of the object or observable in question. It is quite in accordance with the *Hamilton principle* which ultimately produces a description of all classical physics from an *Energy* view of those physical observables. Furthermore the Quanto-Geometric Operator prescribes that there is always a dynamic spatial component to the observable, directly reflected in its eigenform, whether pure or composite, which is the ontological wavefunction component of the observable as modeled by the term:

$$\left[\frac{2}{a_i e^{\left(\frac{2a_i+b_i^2}{4}\right)} + b_i e^{\left(\frac{2a_i+b_i^2}{4}\right)} + ce^{\left(\frac{2a_i+b_i^2}{4}\right)}} \right].$$

Thus far in modern physics, we have been completely oblivious to this representation having maintained a predominant view of pure and quasi exclusive compactification about the ontology of physical observables. For that matter, the above reported quantities from our physics have no accompaniment in the space realm.

Contrastingly, this treatment allows us to capture the universality of the background throughout the span of Shells of matter in our 3-dimensional universe, while preparing us to capture the eigenvalue elements that make up the specific invariants upon which each Shell is built, those which modern physics have denoted the fundamental physical constants of Nature. The visualization of the transverse Shells of matter (Fig.4.11) beyond the important but insufficient Einsteinian concepts of Lorentzian transformations on Minkovski spacetime contemplated in the macroscopic arena, along with the ensuing computation of the pivotal invariants that sustain them, constitutes the royal way that takes us to Grand Unification.

To elicit the application modality of the Quanto-Geometric operators across all Shells of matter is to:

1. Apply to the observable any necessary mathematical transformation that results in turning up its <u>implicit</u> functional expression, if said expression is clearly un-apparent at the outset.

2. Scrutinize the implicit functional expression for conformity to any one of the two ontological tenets of the Quanto-Geometric principle in the form depicted above for the Operator for a direct representation of these forms or thru integra-tion, according to the prescription below:

$$\|QG\| = \left[\left(\int_B^A f \cdot ds \right) \bullet g(s) \right] \equiv \left[-\frac{Q^2}{2q} \right] \bullet \left[\frac{2}{a_i e^{\left(\frac{2a_i + b_i^2}{4} \right)} + b_i e^{\left(\frac{2a_i + b_i^2}{4} \right)} + c e^{\left(\frac{2a_i + b_i^2}{4} \right)}} \right],$$

f being the functional expression for the scalar component and g the wavefunc-tion component.

3. Analyze the coefficients of these terms in comparison with the Quanto-Geometric eigenvalue cosets in order to determine the exact Operator among the set of 9 the observable conforms to.

When it comes to the much celebrated relativistic Energy formula $E = mc^2$, in the light of the above constructs, one might wonder why it did not turn out to be instead $E = mc^2/2$ in straight conformance to the scalar component of the generic Quanto-Geometric Operator.

In attempting a response to this hypothetical, it is important to recognize that the relati-vistic Energy formula does not refer to the wavefunction component of objects but strict-ly to the scalar component and appears as a degeneracy of the latter. Effectively the rela-tivistic energy formula is admittedly a statement of the energy harbored by an object as a result of its mass. Since this expression is also valid for an object at rest $(E = m_0 c^2)$, it is therefore a statement that c must be an attribute of a spatial property participating in the scalar ontology of the object. The best way to interpret this notion is to consider mass or all physical scalar indeed as a density quantity, as posited by the scalar component of the Quanto-Geometric Operator. On that account the expression of the relativistic energy is largely equivalent to the former.

Furthermore, the Relativistic Kinetic Energy expression *(KE)* is going to put us closer to the Quanto-Geometric scalar expression:

$$KE = mc^2 - m_0c^2 \text{ (4-8)},$$

which so defines the relativistic kinetic energy of a particle as energy in excess of its ontological or rest-mass energy. Because there cannot be a conflict between the internal wavefunction of the scalar tenet and the actual explicit wavefunction component of the moving object, there must be a proportionality relationship between the propagation rate associated with the ontological wavefunction in all facets *(c)* and the exogenous speed of the moving object *(v)*. In other words, the moving object can never move at a rate that is larger than the propagation rate of its basis wavefunction (both as part of its scalar packet and its space spread tenet). That is where the Lorentz transformation factor comes into play as well as the specifics relating to the internal quanto-geometric covariance at play. We will not address here the rather involved study of quanto-geometric covariance in this context which is undertaken in Volume II of this publication. We will proceed only by calling the direct ratio between the two speeds as well as the Lorentz factor into consideration. When v is much smaller than c, we have:

$$\text{for } v/c << 1 \rightarrow KE = mc^2 - m_0c^2 \approx \frac{m_0v^2}{2} \text{ (or } \frac{p^2}{2m_0}\text{)}$$

$$\text{For } v/c \rightarrow 1 \rightarrow KE = \frac{1}{2}m_0v^2 + \frac{3}{8}\frac{m_0v^4}{c^2} + \frac{5}{16}\frac{m_0v^6}{c^4} + ... \text{ (4-9)}$$

We arrive at expression *(4-9)* for v approaching c thru the Lorentz conversion factor:

$$\gamma = \sqrt{1 - \left(\frac{v}{c}\right)^2}$$ with its factoring into the Relativistic Kinetic Energy expression, and

additionally thru the binomial expansion of the latter. The preeminent role played by the standard scalar term in both expressions is inarguable and need no further argument.

The above analysis requires us to firmly bear in mind two key points:

- The quanto-geometric scalar is always a dimensional scalar. That is to say that all mass is in the form of density in and of itself.

- When all tensors vanish, including within the dimensional scalar component of an object, the parabolic tensor inalterably persists as a property of the background basis space (Universal 3-space) that participates in the mass component buildout of the object. The single most important dynamic property associated with the parabolic tensor is c.

I will conclude this section by stating that while Lorentzian transformation based on the invariance of c in all reference frames is the foundation of relativistic mechanics, we are going to supersede that vision by directly deriving c from the very axiomatic of the Quanto-Geometric framework, in a demonstration undertaken in the next Chapter.

4.17. Phenomenological Primitives

In this brief survey of analytic structures and methods inherent to the Quanto-Geometric Theory, we shall mention that the class of 9 ontologies in the Functional order, deriving into the orders of Operators and Tensors, has a precise echo in phenomenology. If we assume the ontologies to be Primitive Identities, their strict numeric signatures give rise to behavioral or phenomenological attributes.

We have chosen the nomenclature summarized in Table 4.1 for the Quanto-Geometric Primitives in phenomenology. Barring some variance in the denominations, they are exactly equivalent one to one to the Primitives studied in the previous Chapter. We are only making here a more narrow connection in the nomenclature with self-adjoint or calculational operators. One should read the first column of the Table as Scalar State-1, Scalar State-2, etc. or Primitive-1, Primitive-2 and so on and so forth. The integer symbol is not to mean an ordinal for the purpose of ranking but stands for a scalar state or a Primitive.

This characterization directly applies to all physical observables but is very particularly useful when approaching unknown phenomena initially unyielding to Functional characterization. In the life sciences, or the karyotic span in the 3-D universal arena, where strict and full mathematical formalism may be of difficult application at first, the phenomenological approach is well granted in the attempt to elucidate their dynamics, well beyond the usual statistical accounts. Nevertheless, one must remain fully aware that there is not one single aspect of physical reality that escapes full Quanto-Geometric characterization, as much as there exists no physical observable without a direct numeric expression. Every numeric or scalar state is a direct path to phenomenology and from phenomenology the observer may directly funnel into the deep realms of Functional analysis. All in all, the analyst must be convinced that every physical object is quanto-geometrically instantiated in terms of identity, attribute and behavior (in accordance with the Tri-Valent canon), all three of which make up their ontological signature.

Scalar State/Primitive	Numeric Identity	Phenomenology
→1	Monolithic	Egocentrism
→2	Binary	Dualism
→3	Associative	Relational
→4	Radical	Ambivalence
→5	Imaginary	Metamorphism
→6	Matrixial	Harmony
→7	Irrational	Paradox
→8	Algebraic	Perfection
→9	Transcendental	Infinity

Table 4.1 Nomenclature of Phenomenological Primitives

4.18. Conclusion

In this presentation of the Quanto-Geometric Theory, we have exposed an all-encompassing framework that addresses the nature of space and the quantum-scalar along with their correlation throughout the continuum of the Universal Arena. We have elicited an order of Operators and Tensors directly applicable to the entire continuum of the 3-dimensional Arena. This demonstration represents a decisive step leading to the achievement of theoretical Unification long sought thru the momentous derivation of the entire suite of fundamental physical constants of Nature. In the aggregate, in assessing the value of any system of constructs toward Grand Unification, one must remain keenly aware that the gauging barometer is not at all the proliferation of "beautiful" mathematical tours de force for their own sake but the ostensible ability to derive consistently, effectively and unequivocally from that theoretical framework the fundamental constants of Nature that modern physics has over time arrived at only empirically.

Chapter 5

All things of your ignorance together make up
a realm that is much grander than you.
Proverb by the **Fon-Yoruba People**
of Haiti

DERIVATION OF THE FUNDAMENTAL
PHYSICAL CONSTANTS OF NATURE

In this Chapter we will attempt to put to rest the long debate between the meta-physical view of the Universe contended by occultists and religion in general and the strictly physical view upheld by rationalists such as scientists, mathematicians and others. This discussion posits a framework of formalism from which a whole suite of the most fundamental physical constants are directly derived. It further sets rational grounds for an interpretation of the meta-physical realms from where paranormal experience develops. I must warn the reader however that this discussion requires sound knowledge of intermediate-level mathematics to be amenable, lest the reader is willing to simply lend credence to results and conclusions and overlook the mathematical developments. A sound grasp of the concepts discussed in any of the two previous Chapters is key to an understanding of what is to follow.

5.1. Mappa Mundi of the Universe

The primary Quanto-Geometric function, in its role of universal modeler of all things in existence, represents the grand map of the Universe in its fullness.

$$Q(s) = \pm \left| \frac{s^0}{\sigma \cdot \sqrt{2\pi}} \bullet e^{-s^2/2\sigma^2} \right|, \ s \neq 0$$

The reader recalls the primeval q_n symmetry-breaking sequence introduced in Chapter 2 which helps delineate the suite of natural curvatures embedded in the function:

$$\{q_n\} = \left\{ \frac{e^{(1-n)/(10-n)}}{\sigma\sqrt{2\pi}} \right\}, \text{ where } 1 \leq n \leq 9$$

The quantity n relates to s and determines the *degrees of freedom of space*. By that measure there exist 9 different dimensions of matter, each based on its own fundamental Quantum type. That generic quantum is a scalar object that may be a *quantum* proper as visualized in Quantum Theory or somewhat of a Mass as visualized in classical and relativistic physics. The value of that distributive *quantum* entity defines each Universal Domain as displayed below.

$$Q_1(s) = q_1 = \frac{e^0}{\sigma\sqrt{2\pi}} = \frac{1}{\sigma\sqrt{2\pi}}$$
$$\rightarrow \text{for } n = 1$$

$$Q_5(s) = q_5 = \frac{e^{-4/5}}{\sigma\sqrt{2\pi}}$$
$$\rightarrow \text{for } n = 5$$

$$Q_2(s) = q_2 = \frac{e^{-1/8}}{\sigma\sqrt{2\pi}}$$
$$\rightarrow \text{for } n = 2$$

$$Q_6(s) = q_6 = \frac{e^{-5/4}}{\sigma\sqrt{2\pi}}$$
$$\rightarrow \text{for } n = 6$$

$$Q_3(s) = q_3 = \frac{e^{-2/7}}{\sigma\sqrt{2\pi}}$$
$$\rightarrow \text{for } n = 3$$

$$Q_7(s) = q_7 = \frac{e^{-2}}{\sigma\sqrt{2\pi}}$$
$$\rightarrow \text{for } n = 7$$

$$Q_4(s) = q_4 = \frac{e^{-1/2}}{\sigma\sqrt{2\pi}}$$
$$\rightarrow \text{for } n = 4$$

$$Q_8(s) = q_8 = \frac{e^{-7/2}}{\sigma\sqrt{2\pi}}$$
$$\rightarrow \text{for } n = 8$$

$$Q_9(s) = q_9 = \frac{e^{-8}}{\sigma\sqrt{2\pi}}$$

$$\rightarrow \text{for } n = 9$$

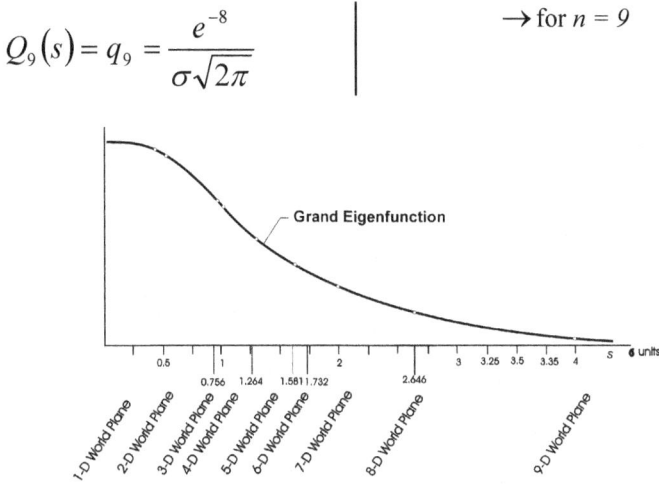

Fig. 5.0 World Planes composing the Omniverse

In other words, the Universe as a whole is made up of 9 distinct domains (Fig. 5.0), each based on a specific *degree of freedom* or *dimensional set*. This characterization sets the basis for the visualization of the universe as an Omniverse with 9 vertically layered domains or World Planes. The above set of relationships may be rewritten to cast a figure of momentum deriving from the Quanto-Geometric articulation inherent to the function. We denote this momentum the *standard* Quanto-Geometric Momentum of a Universal Domain or World Plane. An important point to keep in mind is that the independent variable *s* here is not a simple distance but a *degree of space propagation or dilation,* which ultimately gives rise to distances or lengths. The best physical representation for *s* is what a descriptor for a wave refers to, a *wavelength (or a wavnumber)*. It is a figure that characterizes both motion or propagation and the length associated with a wave phenomenon as a *moving periodic* entity. **This moving periodic entity is physical space itself.** Rewriting the equations above in order to make explicit the momentum figures, we obtain:

$$q_1 \times \sigma = \frac{1}{\sqrt{2\pi}} \text{ , for } n = 1$$

$$q_2 \times \sigma = \frac{e^{-1/8}}{\sqrt{2\pi}} \text{ , for } n = 2$$

$$q_3 \times \sigma = \frac{e^{-2/7}}{\sqrt{2\pi}} \text{ , for } n = 3$$

$$q_4 \times \sigma = \frac{e^{-1/2}}{\sqrt{2\pi}} \text{, for } n = 4$$

$$q_5 \times \sigma = \frac{e^{-4/5}}{\sqrt{2\pi}} \text{, for } n = 5$$

$$q_6 \times \sigma = \frac{e^{-5/4}}{\sqrt{2\pi}} \text{, for } n = 6$$

$$q_7 \times \sigma = \frac{e^{-2}}{\sqrt{2\pi}} \text{, for } n = 7$$

$$q_8 \times \sigma = \frac{e^{-7/2}}{\sqrt{2\pi}} \text{, for } n = 8$$

$$q_9 \times \sigma = \frac{e^{-8}}{\sqrt{2\pi}} \text{, for } n = 9$$

The *standard* momentum, represented by the expression $q_i \times \sigma$, where σ, a value of *s*, is the standard deviation for the Global Universe. It lends its name to the chosen denomination for this figure. In all, q_i is the figure that specifically defines every Universal Domain.

The one-dimensional Universe is given by $Q_1(s)$ or q_1 and is represented in the numerous existing Black Holes and Neutron Stars throughout the cosmos. These entities are knowingly one-dimensional.

The 2-dimensional Universe is given by $Q_2(s)$ or q_2 and is represented in the deployments of Plasmic Matter in Stars and at the center of planets. Plasma, of which fire is a type, is knowingly 2-dimensional.

The 3-dimensional Universe is given by $Q_3(s)$ or q_3 and is represented in what we commonly call *our* physical universe, the visible Cosmos.

The 4-dimensional Universe is given by $Q_4(s)$ or q_4 and represents the first of meta-physical universes. It is a trans-physical universe that cannot fully fall under our observation for reasons we will explain later.

The 5-dimensional Universe to the 9-dimensional Universe are respectively given by $Q_5(s)$ or q_5, $Q_6(s)$ or q_6, $Q_7(s)$ or q_7, $Q_8(s)$ or q_8 and $Q_9(s)$ or q_9, each representing a meta-physical domain that cannot fall under our observation at all because they pertain to the larger post Quanto-Geometric hierarchy. Each of these Universal Domains or World Planes are contained within one another in accordance with the Quanto-Geometric mapping function. While we can observe but not penetrate into the domains that are below us in the order of embedding, it is understandably not possible to observe the ones above which our domain is embedded in.

5.2. Theoretical Derivation of the Speed of Light

How do we know that the above map is correct and not just a speculation? For an answer, let us start by scrutinizing further the value of expression $q_3 \times \sigma$, the standard Quanto-Geometric Momentum for our 3-dimensional universe, by developing the expression in completely rational form:

$$q_3 \times \sigma = \frac{e^{-2/7}}{\sqrt{2\pi}} \quad (5\text{-}0)$$

$$q_3 \times \sigma = 299796065 \times 10^{-9}$$

It is not difficult to recognize in this number the figure or digital set that we know as the speed of light *c: 299,792,458 m/s*. From relation *(4-3)* we have picked the value of the dimensional set for the 3-D World Plane and replace it in *(4-4)* to obtain the s_3 wavenumber which characterizes space in 3-D, that is:

$$\text{For } n = 3, \ s_3 = \pm \frac{2\sigma}{\sqrt{7}}$$

Once we obtain s_3, we replace it in expression *(5-0)* of the Grand Eigenfunction to obtain the value of the Function at that point. So:

$$\text{For } s_3 = \pm\frac{2\sigma}{\sqrt{7}}, \quad Q_3(s) = \frac{e^{-2/7}}{\sigma\sqrt{2\pi}} = \frac{0.299796065}{\sigma}$$

$$\text{or } q_3 \times \sigma = 0.299796065$$

$$\text{or } q_3 = 0.299796065 \cdot \sigma^{-1}$$

In summarized form:

$$\begin{array}{l}
n_3 = 3 = D_3 \\
s_3^{-1} = 1.322875656 \cdot \sigma^{-1} \\
q_3 = 0.299796065 \cdot \sigma^{-1}
\end{array}$$

For the purpose of the demonstration, once we arrive at the 3rd expression we could squarely conclude the analysis. It is possible, however, that a professional physicist might not have been swayed by this result due to sheer unfamiliarity with the expressed quantity for q_3. In proceeding with dimensional-unit analysis for further clarification, our first observation of these results is that, if σ^{-1} is to be taken for a unit of measure, the value of the parabolic scalar and that of reciprocal space faithfully and consistently reflect the principle encapsulated in expression *(4-1)*, namely that nothing exists beyond the two irreducible tenets of matter and that as such they can only be formulated in terms of one another. The system is wonderfully self-contained because we see that the figure of the standard deviation σ arises as a natural endogenous unit of measure within the system, here in its reciprocal dimension.

In our quest for c, let's note that the quantity appearing in the 3rd expression is a clear precursor of the quantity experimentally known to represent c:

$$\mathbf{0.299796065 \, \sigma^{-1}} \rightarrow \mathbf{0.299792458 \cdot 10^9}$$

on account of a large figurate match. This expression gives us the value of the parabolic scalar in reciprocal standard deviation σ^{-1} unit. The reciprocal standard deviation of the Omniverse is a quanto-geometric wavenumber in nature, representing both a length and a propagating wave, that is. We set out to look for a quantity whose known nature gives us a match with the identity expectation that we hold for σ^{-1} or σ, no matter the cardinality or magnitude of the quantity.

This number is indeed an already known quantity in physics in the personification of the *nano-meter*, which is a sub-unit element of the meter scale. The meter being no longer just a unit of measure based on a physical etalon but a fundamental constant, the nanometer sub-element inherits that condition and thus equally stands for a constant in its own right. Most importantly, the nanometer is the metric for radiation wavelengths in spectroscopy, although the BIPM (Bureau International des Poids et Mesures) has given it the general definition of a *unit of length in the metric system equal to 1 billionth of a meter*:

$$1nm = 1.00000\overline{0} \times 10^{-9} m$$

The metrology of this unit is typically utilized in condensed matter physics and below, very particularly in the field of spectroscopy, for the measurement of **light** radiation (not a coincidence) resulting from energy transitions of particles in atoms and molecules. Therefore the match in the "dimensional" nature of the *nanometer* constant and the nature of the standard deviation figure σ is a solid given and is completely natural. We shall add as a note that the Angstrom unit (Å) with the value of *0.1 nanometer* has also been in use in spectroscopy but has been fairly superseded by the *nanometer* due to better suitability to task.

Our motivation being the derivation of the <u>exact</u> value known for c, we are going to allow ourselves to fine-tune ever so slightly the value of the *nanometer* to the following magnitude:

$$1nm = 1.00001203 \times 10^{-9} m$$

We may well do so because the official value of *1.0 x 10^{-9}m* of the nanometer, as a dependency on the meter, is not known to be absolutely exact and a change in the order of the 5th decimal place is, from that standpoint, absolutely inconsequential to the applicability of the sub-unit. Furthermore, and most importantly, we primarily port this adjustment in magnitude to σ, an element in our own system, so that:

$$\sigma = 1.00001203 \times 10^{-9} m$$

We thus register the value of the standard deviation of the Omniverse in its reciprocal form to be exactly the above reported quantity. By substitution of σ^{-1} in the parabolic scalar relation, we obtain:

$$q_3 = \frac{0.299796065}{1.00001203} \times 10^9 \quad \text{or}$$

$$\boxed{q_3 = 299\,,792\,,458}$$

exact magnitude of the speed of light! This quantity, however, is an atemporal figure! It is important to recognize that we have not arrived at this quantity by introduction of speeds or ratios of *m/s* in the algebra, but strictly by a legitimate manipulation of the metric, and in complete compliance to the time-free condition of the framework. This quantity is the measure of the *Moment of Propagation* of the parabolic *eigenwave* that makes up dimensional background space in the 3-D World Plane. The wavelength of the propagating *eigenwave* is given in the value of reciprocal space $1/s_3$. In other words, *c* is a *quantity of motion*, the motion *inherent* to background *3*-space in the Omniverse. All the transversal 3-D Shells of matter, from the quark scale to the large-scale cosmic structures, have developed over and from this spatial background, and all physical motion occurs by degeneracy from this dynamical fabric. All manifolds develop from this dynamical fabric as well and that is why at any point in the fabric when all tensors vanish the parabolic *eigenform*, otherwise called the Einstein tensor, indefectibly subsists. The latter is however matter for another analysis. By the same token, this analysis patently explains why *c* as the moment of propagation of the basis spatial eigenwave of the 3-D World Plane cannot be surpassed as long as we remain within the boundaries of the 3-D realm.

If the value of the physical constant *c* is indeed *299,792,458*, then the value of the standard deviation of the Omniverse *(σ)* of *1.00001203 x 10*$^{-9}$ must consequently be an outright true physical quantity in its own right (not just a mathematical scalar) and an effectual one as well. The best way to interpret this result is to visualize the parabolic scalar quantity *(q₃)* of **0.299796065σ**$^{-1}$ as a pure mathematical quantity, while the same quantity expressed as the scalar **0.299792458·10**^{9} is a mathematical-physics quantity that comes into existence when we give physical meaning to the standard deviation scalar. This all means that the meter etalon, originally devised by BIPM as a certain distance between two marks on an iridium bar held at a certain temperature, holds an incredible level of accuracy and is true to nature in a highly accurate manner.

The theoretical derivation of the quantity representing the speed of light (or its moment of propagation) stands for the sole and exclusive measure of theoretic support a-posteriori to the decision by BIPM to set that quantity as the standard for the definition of the meter, because it removes arbitrariness in a very large measure from the metric standard. In addition, if the quantity of *1.00001203 x 10*$^{-9}$*m* is found to be effectual in experimentation or observation in the predicted condition, then we would have realized complete synchrony between human theory and the physics of nature under the auspices of a mathematical-physics system, the Quanto-Geometric system! We will come back to this number in Chapter 6 in order to elucidate another important dimension of this quantity as regards to the cosmological expansion of the universal arena.

5.3. Misadventures of c in Time

The table below shows the history of measurements of the speed of light since the 17th century to date. Ever since Albert Einstein's Theory of Relativity became officially accepted in the first half of the 20th century, physicists recognized the towering importance of the speed of light in our understanding of physical reality, and endeavored to measure it with as much accuracy as could be possible. The latest measurements employed the best technological means available such as coherent laser beams.

Nevertheless the task of accurately measuring the speed of light proved to be somewhat elusive to the standard bodies as they recognized that such accuracy depended on the accuracy with which we measure both distance and time, notably the latter. The measurement of time has had its own shortcomings, references for the time unit, the second, having had to be abandoned one after another due their inconstancy, to the point that currently two different scales of time are in use in the United States of America. The standard bodies had even adopted in the past a time unit pegged to a set date in time (year 1900) in order to remain clear of the variations that were naturally occurring in etalons and measurement conditions. It soon became clear that that measure was equally unsustainable.

As a consequence of these shortcomings, the standard bodies, CGPM (Conférence Générale des Poids et Mesures) and NIST (National Institute of Science and Technology) finally decided in 1975 to establish the best measurement taken then of light's travel speed as a reference in itself, and make the unit distance, the meter, and the unit of time, the second, secondary constants which were defined by the very speed of light. Consequently, the speed of light constant has been conventionally set to be *exact*. Now one may wonder what happened to the universality of time, which was established in Newtonian mechanics as "universal, uniform and true", and which has survived to this day thanks to Einstein's constructs of Relativity? It is obvious that the notion of time, just as the notion of distance, has preponderance over the notion of speed of any physical object or artifact. Making time, acknowledged thus far to be a dimension in itself, dependent on c seems to be a counter-intuitive proposition that amounts to putting the cart before the horse.

DATE	INVESTIGATOR	METHOD	MEASURED SPEED (m/s)
1676	Ole Roemer	Jupiter's Moons	220,000
1726	James Bradley	Stellar Aberration	301,000
1834	Charles Wheatstone	Rotating Mirror	402,336
1849	Armand Fizeau	Rotating Wheel	315,000
1862	Leon Foucault	Rotating Mirror	298,000
1868	James Clerk Maxwell	Theoretical Calculations	284,000
1875	Marie-Alfred Cornu	Rotating Mirror	299,990
1879	Albert Michelson	Rotating Mirror	299,910
1888	Heinrich Rudolf Hertz	Electromagnetic Radiation	300,000
1889	Edward Bennett Rosa	Electrical Measurements	300,000
1890's	Henry Rowland	Spectroscopy	301,800
1907	Edward Bennett Rosa and Noah Dorsey	Electrical Measurements	299,788
1923	Andre Mercier	Electrical Measurements	299,795
1926	Albert Michelson	Rotating Mirror (Interferometer)	299,798
1928	August Karolus and Otto Mittelstaedt	Kerr Cell Shutter	299,778
1932 to 1935	Michelson and Pease	Rotating Mirror (Interferometer)	299,774
1947	Louis Essen	Cavity Resonator	299,792
1949	Carl I. Aslakson	Shoran Radar	299,792.40
1951	Keith Davy Froome	Radio Interferometer	299,792.75
1973	Kenneth M. Evenson	Laser	299,792.4570
1978	Peter Woods and Colleagues	Laser	299,792.4588

Table 5.1 Historical measurements of the speed of light c

Well aware of this inconsistency, physicists have justifiably sought to make c as much an independent constant as possible by turning it into an abstraction, thereby giving it a life of its own. From Maxwell theory of electromagnetism, it ensued that:

$$c = \frac{1}{\sqrt{\mu_0 \varepsilon_0}}$$

where μ_0 is the magnetic permeability of space and ε_0 the electric permittivity of space. The values assigned to μ_0, and subsequently that of ε_0, were conventionally chosen in order to satisfy the above equation with the *exact* value that was established for c. Therefore non physicist readers should not be disabused when they see these two constants in Table of Physical Constants published in the literature as *exact* constants. They are rather *conventionally exact*. What's more, these two constants do not describe, and are not intended to describe, any physical property of space whatsoever. Shall we say that the Quanto-Geometric architecture provides the most natural derivation of c from pure abstraction there is.

In accordance with its preeminence within the hierarchy of physical constants, the speed of light constant has become a ubiquitous entity in modern physics, appearing as a parameter in many contexts unrelated to light phenomenon per se. The Lorentz transformation equations are the first of such examples where speeds of objects are formulated in terms of c. General Relativity predicts the propagation speed of gravitation to be c, although gravitational waves have yet to be observed (as of first publication date of this text). The constant c is not only the speed of light in vacuum but generically the speed of all "mass-less" particles in vacuum. All of this clearly means to tell us that it is not about c, but about space. We remain blind to those important facts because we are unable to formulate a model of space as the mother of all motion and we are determined to give fit and life to c at all costs in a world of time of our creation which it does not belong to. Again the Quanto-Geometric framework has been able to properly authenticate and quantify c by derivation because it has established the necessary foundation for that lacking space model as implied by the behavior of c. We must once and for all get rid of the false notion of time to visualize the inherent properties of space. And yes, *spacetime* is the wrong notion.

Individually we are all innately cognizant of the fact that the universe around us is a 3-dimensional domain, that is to say it is endowed with 3 isotropic dimensions of space: length, width and height. A dimension of time is nowhere to be found, except in our human minds as a counter-productive abstraction. These 3 dimensions of space are what ultimately explains the specific value of c and its insurmountable stature in our universe.

5.4. World Planes and Space Propagation

Just as we have seen that the Quanto-Geometric momentum $q_3 \times \sigma$ or $Q_3(s)$ remits to a constant for space propagation in the 3-dimensional Universe, the other momenta of the Mappa Mundi Function remit to specific levels of space propagation in each of the other 8 Universal Domains (World Planes) as well. As we go up the hierarchy of Universes from U_1 to U_9, the value of the standard Momentum should increase and we would expect the suite of equations to reflect an increasing *degree of space propagation*, in their comparable figure to c, from U_1 to U_9. That may not be quite apparent however, unless we call for the *generic* Quanto-Geometric momentum and its specific values from U_1 to U_9. Below is the derivation of all q_i from U_1 to U_9. We use the physical nanometer quantity for σ in the computations.

$$\text{For } n = 1,\ q_1 \times \sigma = \frac{1}{\sqrt{2\pi}} = 0.39894228$$

$$\text{and } q_1 = \frac{1}{\sigma\sqrt{2\pi}} = 398937481.2$$

For n = 2, $q_2 \times \sigma = \dfrac{e^{-1/8}}{\sqrt{2\pi}} = 0.352065327$

and $q_2 = \dfrac{e^{-1/8}}{\sigma\sqrt{2\pi}} = 352061091.5$

For **n = 3**, $q_3 \times \sigma = \dfrac{e^{-2/7}}{\sqrt{2\pi}} = 0.299796065$

and $q_3 = \dfrac{e^{-2/7}}{\sigma\sqrt{2\pi}} = 299792458$

For n = 4, $q_4 \times \sigma = \dfrac{e^{-1/2}}{\sqrt{2\pi}} = 0.241970725$

and $q_4 = \dfrac{e^{-1/2}}{\sigma\sqrt{2\pi}} = 2419677813.6$

For n = 5, $q_5 \times \sigma = \dfrac{e^{-4/5}}{\sqrt{2\pi}} = 0.179256322$

and $q_5 = \dfrac{e^{-4/5}}{\sigma\sqrt{2\pi}} = 179254165.2$

For n = 6, $q_6 \times \sigma = \dfrac{e^{-5/4}}{\sqrt{2\pi}} = 0.114298877$

and $q_6 = \dfrac{e^{-5/4}}{\sigma\sqrt{2\pi}} = 114297502$

$$\text{For } n = 7, \quad q_7 \times \sigma = \frac{e^{-2}}{\sqrt{2\pi}} = 0.053990967$$

$$\text{and } q_7 = \frac{e^{-2}}{\sigma\sqrt{2\pi}} = 53990317.01$$

$$\text{For } n = 8, \quad q_8 \times \sigma = \frac{e^{-7/2}}{\sqrt{2\pi}} = 0.012047013$$

$$\text{and } q_8 = \frac{e^{-7/2}}{\sigma\sqrt{2\pi}} = 12046868.08$$

$$\text{For } n = 9, \quad q_9 \times \sigma = \frac{e^{-8}}{\sqrt{2\pi}} = 0.00013383$$

$$\text{and } q_9 = \frac{e^{-8}}{\sigma\sqrt{2\pi}} = 133828.61$$

This list gives us a decreasing suite of numbers from q_1 to q_9. If we recall the expression for the *generic* Quanto-Geometric Momentum:

$$q \times s = 1,$$

$$\text{then } s = \frac{1}{q}$$

$$\text{or } s_i = \frac{1}{q_i}$$

Thus the value for each of these degrees *(s$_i$)* of space propagation is the inverse of the momentum *(q$_i$)*, which sets them to be on an increasing slope from U_1 to U_9. In other words, the maximum attainable speed in each Universe of the Omniverse is higher from one to the next without ever reaching either infinite (instantaneity) at U_9 or 0 (absolute immobility) at U_1.

Furthermore, there is a distinct quality of the meta-physical universes (from U_4 to U_9) as compared to the physical universes (U_1 to U_3). The fact that we, as observers from the 3-dimensional realm, cannot perceive them at all is not a trivial fact and testifies to their distinct quality in terms of their Quanto-Geometric momenta. We can only become observers to these other World Planes by using technological transforms such as based on the superheterodyne principle but would never be able to physically penetrate them. We will later see what that level of detection consists in. All of the Quanto-Geometric principles fully apply here, including the signatures attributable to each layer as qualifiers and descriptors for each of these Universes individually. It is worth adding that although we may have an idea of what space may be like in U_4 to U_9 thru the cardinality of their different dimensions, we are and will remain completely unable to conceptualize what the 4-th, 5-th ... 8-th and 9-th dimension really mean from the boundaries of 3-dimensional abstractions. Theosophy gives a path to that knowledge which, however, implies a completely different vision of ourselves as living entities, a matter that we shall address later in this work. As elusive as these meta-physical universes might be to us, they remain nevertheless completely real domains, as real as our 3-dimensional Universe, in accordance to their signature within the Quanto-Geometric framework.

5.5. Horizontal Hierarchy of Domain Shells

Let us start by positing the following relationship, which equates the exponent of e in the Grand Mappa Mundi function of one variable to the primeval sequence:

$$-\frac{s^2}{2\sigma^2} = \frac{1-n}{10-n}, \text{ where } 1 \le n \le 9$$

$$s^2 = -2\sigma^2 \frac{1-n}{10-n}, \quad s^2 = \sigma^2 \frac{-2(1-n)}{10-n}$$

$$s^2 = -\sigma^2 \frac{2(1-n)}{10-n}, \quad s^2 = \sigma^2 \frac{2n-2}{10-n}$$

$$s^2 = \pm \sqrt{\frac{\sigma^2(2n-2)}{10-n}}$$

$$s = \pm\sigma\sqrt{\frac{(2n-2)}{10-n}}$$

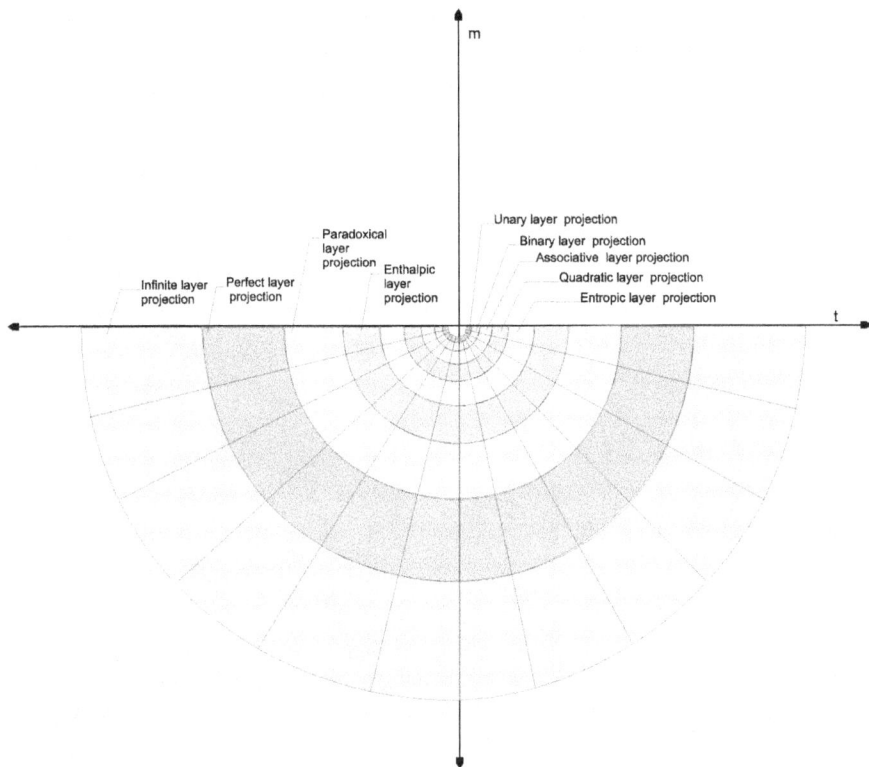

Fig. 5.1 Projection of the graph of Quanto-Geometric function on the *m-t* plane
displaying the normalized groups

In other words, to every individual Universe in the vertical Hierarchy, going from U_1 to U_9 and in accordance with their respective generic Momentum, corresponds an interval of matching s values. We know, from formerly discussed Quanto-Geometric principles, that there are 9 critical s values, by which the $Q(s)$ function can be horizontally normalized. To find those 9 critical values of s, we may replace integer n by an allowed value within its allocated range, in the last of above equations. After performing those replacements, the list of corresponding s values are as follows:

for $n = 1$, $s_1 = 0$

for $n = 2$, $s_2 = \pm\sigma/2$

for $n = 3$, $s_3 = \pm \sigma 2/7^{1/2}$

for $n = 4$, $s_4 = \pm \sigma$

for $n = 5$, $s_5 = \pm \sigma(8/5)^{1/2}$

for $n = 6$, $s_6 = \pm \sigma(5/2)^{1/2}$

for $n = 7$, $s_7 = \pm 2\sigma$

for $n = 8$, $s_8 = \pm \sigma 7^{1/2}$

for $n = 9$, $s_9 = \pm 4\sigma$

Again it cannot be overstated that these values of s mean more than just a distance in space but also a *form* of space. They stand for *wavenumbers*, derivates of wave-lengths which characterize both a distance and a *degree of propagation* of space. From one interval to the next, space experiences more freedom or more extensiveness (longer wavelength) resulting in a larger capsule or a larger domain as we go down the line. We recall from Chapter 2 that the list of space points reported above emanate from a primeval symmetry-breaking sequence. So therefore the different n degrees of freedom of physical space yield the following space Intervals for the whole Omniverse:

Interval 1 $\rightarrow$]0, $\sigma/2$[, for $n = 1$

Interval 2 $\rightarrow$ [$\sigma/2$, $2\sigma/7^{1/2}$[, for $n = 2$

Interval 3 $\rightarrow$ [$2\sigma/7^{1/2}$, σ[, for $n = 3$

Interval 4 $\rightarrow$ [σ, $\sigma(8/5)^{1/2}$[, for $n = 4$

Interval 5 $\rightarrow$ [$\sigma(8/5)^{1/2}$, $\sigma(5/2)^{1/2}$[, for $n = 5$

Interval 6 $\rightarrow$ [$\sigma(5/2)^{1/2}$, 2σ[, for $n = 6$

Interval 7 → [2σ, σ7$^{1/2}$[, for *n = 7*

Interval 8 → [σ7$^{1/2}$, 4σ[, for *n = 8*

Interval 9 → [4σ, ∞[, for *n = 9*

The rotation of the Quanto-Geometric function of one variable around the q-axis in order to recover its full 3-dimensional layout is going to transversally spread these intervals over circular spans around the q-axis as depicted in Fig. 5.1. By that account alone, each universe would have a single uniform type of space and a single class of scalar unit (quantum/mass), as it pertains to the postulated generic Quanto-Geometric momentum. However our experience in the 3-dimensional universe tells us that it is not quite so: the universe around us is a deployment of space capsules. This is because in order to generate an individual physical Universe, the *work of creation* puts forth a transform and an operator, the nature of which we shall shortly discuss.

5.6. Operators and Transforms of Creation

The work of creation instantiates 3 quintessential Moments under the regency of the Tri-Valent Quanto-Geometric Operator. The first instantiation originates from the Hyper Quantum Operator and puts forth Uni-Form Operation per its defined norm. The second instantiation is an inheritance of the Hyper Quadratic Operator and puts forth Distributive Operation according to its norm of Ambi-valence. The third instantiation takes origin from the Hyper Space Operator and puts forth the Operation of Space Overture in accordance with its norm of *spatiation* and *spatialization*. The Operator treatment that follows requires us to keep in mind that all Quanto-Geometric operators are self-adjoint operators.

5.7. Hyper Quantum Tri-Valent Operator and the Laws of Conservation

Space in U_3 is characterized as a three-dimensional form of space owing to its origin of parabolic symmetry. The embedded parabolic conic section within the geometry of the Quanto-Geometric function is its native realm. Within the work of creation of U_3, the Uni-Form Operator acts on the dimensional configuration of U_3 space in order to further modulate it, notwithstanding the constancy of its dimensional modulus, which remains inalterably 3.

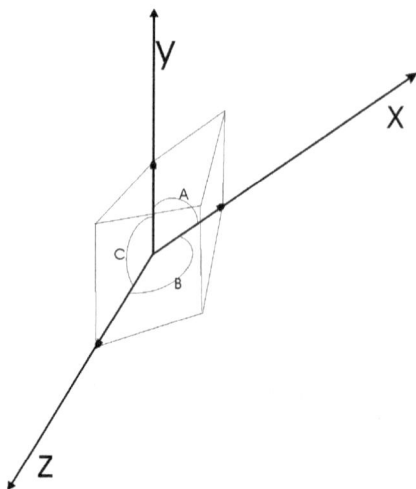

Fig. 5.2 Anisotropic U_3 space in the absence of the Hyper Quantum Operator

Were it not for the intervention of this operator however, the 3 dimensions of U_3 space would be irregular or *an-isotropic* as depicted in Fig. 5.2. (*Tropism* means direction, *iso* means same, *an-* means non). While two of the dimensions would necessarily fall in the same plane, the third dimension would be cast in any irregular direction and the angle they all make with one another could be any angle of any value. Futhermore the unit-length on each dimensional axis would be different, as displayed in Fig. 5.2.

Uni-Form operation by the Hyper Quantum Operator, whose signature is type 1 symmetry or mono-centric symmetry as studied in Quanto-Geometric Number Theory, compells the three dimensions of U_3 space to exhibit mono-centric symmetry, a constraint whereby all directions must exhibit same unit-length, and all angles between them must lay equal to one another. The only possible configuration that responds to this constraint is U_3 space cast from an ortho-normed Euclidean coordinate system, as shown in Fig. 5.3.

The Hyper Quantum Operator thereby constrains U_3 space to demonstrate complete *isotropic* properties and to develop from a perfectly regular cube of action. In that manner, the intervention of the Hyper Quantum Operator in the work of creation is responsible for all the known **laws of conservation** in our 3-dimensional universe thru the type 1 symmetry constraints imposed on the configuration of space in U_3.

1. It is responsible for the *conservation of linear momentum*, a law whereby any object thrown in pure vacuum at a constant velocity is going to keep moving at that same velocity and along the same direction forever.

2. It is responsible for the *conservation of angular momentum*, a physical law whereby any object thrown in pure vacuum at a constant angular velocity is going to keep moving maintaining that same angle of progression forever.

3. It is responsible for the *conservation of energy*, a law whereby any process of a certain type containing a certain amount of energy can change to another process of a different type with the total amount of energy present in the initial state still fully present in the final state after the conversion.

4. It is responsible for the *instantiation and allocation of all masses* to all elemental objects in U_3, in particular the sub-components of particles as well as particles themselves.

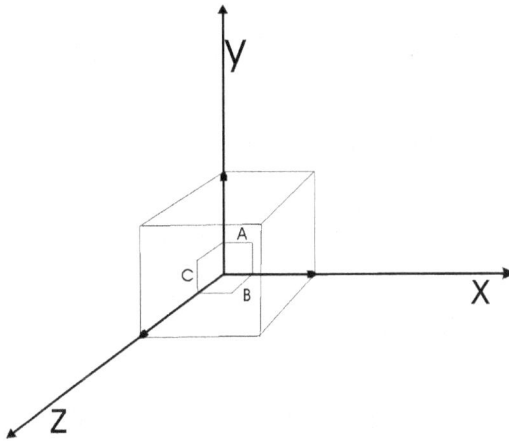

Fig. 5.3 Spatial cube of action in U_3

Were it not for the intervention of the Hyper Quantum Operator, an object thrown at a certain velocity in the purest of vacuum could end up coming to a stop or increasing its travel velocity in and of itself. Absent the Hyper Quantum Operator, an object launched in purest vacuum from a circular movement could end up moving instead according to an ellipse and no longer a circle, or not even in the same plane of rotation, this absent any other intervening factor. Were it not for the Hyper Quantum Operator, no energy transformation process in U_3 could take place without always ending in either a loss or a gain of energy according to the direction along which the process takes place. Absent the isotropic constraints imposed by the Hyper Quantum Operator, and because of the eventual resulting imbalance between dimensional directions in U_3 in that condition, no adiabatic process would most probably be feasible in our universe either.

The instantiation and allocation of masses making up the foundation of U_3 by the Hyper Quantum Operation we shall address in the Volume II of this publication. We will only anticipate here that it is based on a Transform T_1 such that:

$$T_1 \rightarrow \frac{d[Q(s)]}{ds}, \text{ where } s \rightarrow 0$$

the parameter s setting the spin (rotational wavefunction) of the scalar element.

5.8. Derivation of the Planck Constant and the Stephan-Boltzmann Constant

Hyper Quadratic Operator takes two distinct actions in the generation of U_3. Firstly it instantiates a transform, which is the derivative of the Grand Mappa Mundi function, taken as an image of the entire 9-plane Omniverse, and utilizes that image as a sub-map to drive the generation of sub-universe U_3. Secondly it selects two specific values of that function from two particular s points in order to create the first two physical pillars upon which the edifice of U_3 is called to rest: the physical constant of potential energy density and the physical constant of kinetic energy density, namely the *Plank constant h* and the *Stefan-Boltzmann constant* σ respectively. This σ symbol is not the same as the standard deviation s value of the Quanto-Geometric function. We shall designate in this exposition the Stephan-Boltzmann constant σ* rather in order to distinguish it from the Quanto-Geometric standard deviation symbol. The Hyper Quadratic Operator sets the scale and dynamics of quantization, the basis for all forms of segregation and integration in the composition of U_3. By that norm, two elements are to preside over this form of distribution, *Thermodynamic Temperature* and *Energy Quanta*.

Therefore the Transform T_4 such that:

$$T_4 \rightarrow \frac{d[Q(s)]}{ds}, \text{ where } s = (½ \, \sigma, \sigma)$$

represents the Hyper Quadratic Operator's product put forth for the configuration of the 3-dimensional universe. The choice of the two mentioned values of s for the transform is far and away from being arbitrary. We recall that the standard deviation point σ is a critical s value where the derivative of the Quanto-Geometric function or its differential experience shows unique behavior: at that point the first-order derivative function reaches its highest value and the second-order derivative function $d^2Q(s)/ds^2$ is null.

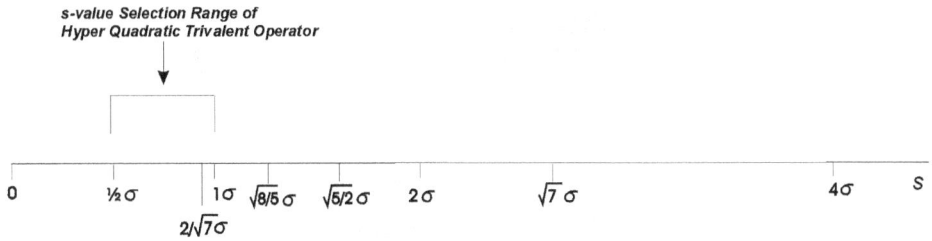

Fig. 5.4 Selection range of s values for the T_4 Transform instantiated by Hyper Quadratic Trivalent Operator

It is all the more natural that this point is selected for the Transform by a Hyper *Quadratic* Operator, which already has a defining *quintessential* life set at that point (Fig. 5.4). Secondly, it stands all the more natural that an Operator whose norm is *Ambi-Valence* instantiates *two* points of action to develop its operation. And finally it stands no less to reason and conformity that, given a choice, this second point be ½ as the one half of the two ambi-valent tenets of the articulation. Therefore even though the point $2/7^{1/2}$ belongs within the range of selectable values, it is summarily discarded by the Operator.

In further detailing the analysis of the transformation, the Hyper Quadratic Operator notably forces the original trinary object into equalized behavior by its two components Q_3 and S_3 according to the very norm of Ambi-Valent action by the Operator. By that norm, Q_3 is forced to induce more space into its own makeup, becoming thereby more distributive and S_3 is forced to borrow from or acquire Q_3's scalarity, inducing tensors into its own texture. This co-valent interaction develops within the boundaries of the primary or generic Quanto-Geometric Momentum of U_3 objects whereby $Q_3 \times S_3 = 1$, a principle that cannot be violated. The ultimate consequence of these constraints is that the new Q_3 and S_3 components of the U_3 unit object acquire comparable equity with one another, although not becoming identical in nature. This is well captured in the popular Eastern ying-yang symbol (Fig. 5.5), a cryptogram that had understandably long fascinated Niels Bohr, one of the lead proponents of Quantum Mechanics, world of conjugates.

Fig. 5.5 Covalent interaction between 3-D scalar and space

They each come to exhibit a Momentum of sorts because they both now have significant space/scalar constitution which is further equalized. This Momentum figure for Q_3 may be expressed as Potential Energy, or to be more exact, Potential Energy Density (because of the derivative backdrop) and its mathematical expression cast as:

$$E_p = h \cdot v \quad (5\text{-}1)$$

where h is the distributive scalar and v the acquired space wave function. Likewise, the Momentum figure for S_3 may be expressed as Kinetic Energy density and its numeric definition cast as:

$$E_k = \sigma^* \cdot T \quad (5\text{-}2)$$

where σ^* is the acquired scalar and T what the principal wave function has become (*tensored* space). The above transforms set the stage for many known transformations in classical energy physics and quantum electrodynamics, notably the reversibility between mass and energy as well as the reciprocal conversion of potential energy and kinetic energy. Let us now take a look at the numbers involved in these relations by starting with the first-order derivative expression of $Q(s)$. When developed the first-order derivative of $Q(s)$ shows:

$$\frac{dQ(s)}{ds} = \frac{1}{\sigma\sqrt{2\pi}} \cdot \frac{-2s}{2\sigma^2} \cdot e^{\frac{-s^2}{2\sigma^2}}$$

$$\frac{dQ(s)}{ds} = \frac{-s}{\sigma^3\sqrt{2\pi}} \cdot e^{\frac{-s^2}{2\sigma^2}}$$

At $s = \sigma/2$, $\quad \dfrac{dQ(s)}{ds} = \dfrac{0.17603266}{\sigma^2} \quad$ and $\quad \left(\dfrac{dQ(s)}{ds}\right)^{-1} = 5.680763 \cdot \sigma^2$

At $s = \sigma$, $\quad \dfrac{dQ(s)}{ds} = \dfrac{0.2419707245}{\sigma^2} \quad$ and $\quad \left(\dfrac{dQ(s)}{ds}\right)^{-1} = 0.413273135 \cdot \sigma^2$

In the above relations, σ is the Quanto-Geometric standard deviation parameter. The official value of the Stephan-Boltzmann constant is: **$5.6705 \cdot 10^1$ W m^{-2} K^{-4}**. And the official value for the Plank Constant is: **0.4135666751 10^{-16} eV·s**. Both of these values are in a satisfactory match with our values for the inverse derivatives above, except of course for the units used, as per determination of the transforms previously discussed. The first relations $(5-1)$ is well known in Quantum Electrodynamics and indeed there is equivalence between h in that equation, standing as the distributive scalar of the unit object in U_3, and the Planck's quantum of action, the smallest energy packet in U_3. In passing, one should know that the Planck Constant is admittedly the fundamental physical constant measured to the worst level of precision. What may be new territory are the implications of the second relation $(5-2)$, only a variation of which is known. Let us compare that relation to the Stephan-Boltzmann law:

$$j^* = \sigma^* \cdot T^4 \quad (5\text{-}3)$$

where j^* is the emissive power or energy flux emanating from an irradiated black body and T the thermodynamic temperature. We can see that the underlying notion subtended by j^* is a fair match to the notion of kinetic energy density, the only apparent discrepancy relating to the 4th power to which T is elevated. Is it any surprise however that a *Quadratic* operator, with numeric definition emanating from the set of power numbers on the real number line, establishes a relationship based on a quantity not only elevated to a power but additionally a power of 4? In that relation T is expressed in Kelvin. If we make $T^4 = N$, the relations becomes:

$$j^* = \sigma^* \cdot N \quad (5\text{-}4),$$

where N is any real number elevated to the 4th power comprised within the Kelvin scale of temperature. This equivalency gives us a match, under the condition that N must be a number representing a power of 4 within the boundaries of the Kelvin scale ($N = T^{1/4}$). We are consequently driven to three conclusions, namely that:

1. The 4th power of the Kelvin scale of temperature is the *natural* scale of temperature (K^4).

2. Temperature is a physical variable that represents native kinetic energy. This is in fair agreement with classical definition of Temperature, which is cast as the average kinetic energy of all unit objects in a population of such objects.

3. The physical constant σ^* (the Stephan-Boltzmann constant) represents the smallest possible unit or "degree" of heat or *quantum* of heat in U_3 as it were.

	Spinor space → $s_i = \frac{1}{2}\sigma$		Spinor space → $s_i = 1\sigma$	
dq/ds	$0.1760\,3266/\,\sigma^2$	Electron Gyromagnetic Moment	$0.2419707245/\,\sigma^2$	Electron Compton Wavelength
$(dq/ds)^{-1}$	$5.680763^+\,\sigma^2$	Stefan-Boltzman constant	$0.413273135^+\,\sigma^2$	Plank constant (in eV s)

Table 5.2 High-level black-body-related physical constants

Please refer to the Table of high-level constants above (Table 5.2) as well as Table 5.6 of Physical Constants further down in this Chapter, offered in support of this analysis of the theoretical derivation of known physical constants. Also offered in Appendix I for reference is the complete repertory of known physical constants with their current official values released by CODATA 2010 (Committee for Data on Science and Technology by the United States' National Institute of Science and Technology).

We remind the reader that a black body is by definition an idealized object able to absorb all the radiation it receives. The interesting fact about such an object is three-fold:

- At a certain point the distribution of energy throughout the different wavelengths of the radiation takes the form of normalized distribution (incidentally and principally the form of the derivative of our image function) independently of the shape of this body.

- At that point the temperature of the body has stabilized and is the same everywhere throughout the emissive object or emissive part of the object as well as the radiation spectrum.

- The total amount of energy distributed over the radiation wavelengths is dependent only on the uniform equilibrium temperature.

The salient fact to take into account from the above is probably not the radiation energetic spectrum or the shape or constitution of the body that is home to the interaction, but the *uniformization* of temperature when the interaction between the radiation and the black body stabilizes and that the temperature level alone is the single most significant dependency at the state of equilibrium (Fig. 5.6). This is captured in the known thermodynamic relation *(5-3)* above.

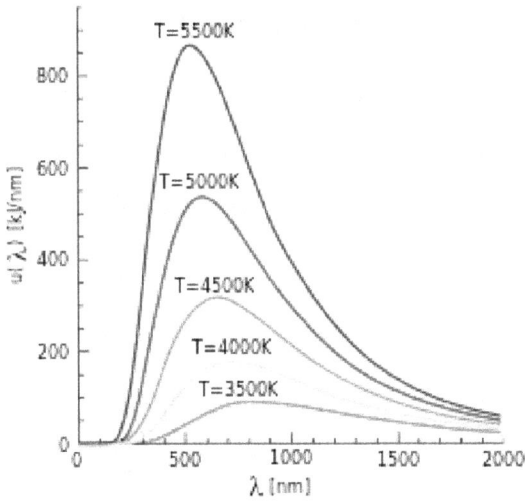

Fig. 5.6 Black body radiation spectrum at different temperatures

An interpretation of this formula within our framework sets σ^*, the Stephan-Boltzmann constant, to be a constant of proportionality for *Temperature* and that the smallest quantity or unit temperature that will ever make sense is σ^*. We shall remember that the Temperature of a macroscopic object is defined as the average kinetic energy of the molecules making up that body. In fact Temperature is the average kinetic energy of all unit-scalar elements making up an object, whether the object is a population of particles, a population of atoms, molecules, etc. So therefore, Quanto-Geometrically, the Stephan-Boltzmann constant in the portrait of:

$$\left(\frac{dq}{ds}\right)^{-1} = 5.680763 \cdot \sigma^2,$$

is what we might call in its own right the *Temperature constant*. The Stephan-Boltzmann constant, officially measured to be **$5.6705 \cdot 10^1$ W m^{-2} K^{-4}**, duly shows the temperature unit in the dimensions of the measured value of the constant in K^{-4}. So it is because it expresses the recognition that the constant constitutes the very etalon for temperature or *thermo-dynamic* temperature. By the same token we now also know why the radiation spectrum of hot objects has the peculiar form that it takes as shown in Fig 5.6.

X-Ray Radiation Spectra

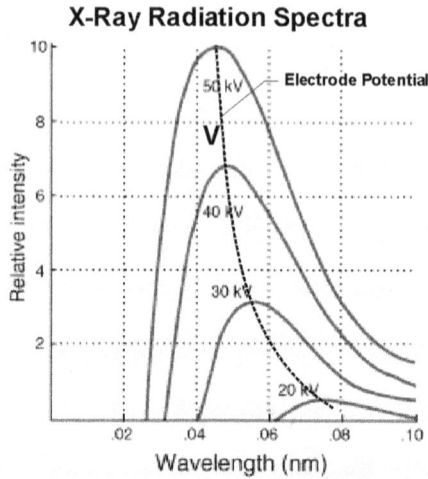

Fig. 5.7 X-ray radiation patterns at different anode-cathode voltages

Effectively one thing is to hold a formula that describes the distribution of energy across a radiation spectrum, and a totally different other of no less significance is to elucidate why the distribution has the specific form or geometry that it portrays and not any other. We know that the geometry of this distribution is principally an inheritance of the geometry of the derivative germane to the generating image function of the Mappa Mundi which is provisioning the entire physical architecture of U_3. To that extent, another radiation distribution pattern, that of X-rays, may become of interest in that it shows a similar pattern (Fig. 5.7).

The uniformizing element of thermodynamic temperature is hereby replaced by the steady voltage between electrodes within the evacuated X-ray chamber. Nevertheless the distribution follows the same pattern as that of black body radiation of visible light both in the geometry of every individual distribution spectrum and in their displacement as a function of voltage.

5.9. Derivation of the Electron Gyromagnetic Ratio and the Electron Compton Wavelength

One may inadvertently overlook the other two quantities mentioned above, the two straight values of $dQ(s)/ds$, namely:

$$\frac{0.17603266}{\sigma^2} \text{ and } \frac{0.2419707245}{\sigma^2}.$$

The first is indeed what we know as the *Electron Gyromagnetic Ratio* with the official value of **0.176085974 · 10¹⁰ Rad s⁻¹ T⁻¹**, and the second is the *Electron Compton Wavelength*, with the current official value of **0.242631058 · 10⁻¹¹ m**.

The *Electron Compton Wavelength* is the equivalent to the wavelength of a photon whose energy is the same as the rest-mass of the electron. The interaction that yields this quantity is the collision between an incident photon and the free electron, which scatters away or recoils after the collision. The Compton wavelength for the electron is given by the formula:

$$\lambda_e = \frac{h}{m_e c}$$

where h is the Plank Constant, m_e the rest-mass of the electron and c the speed of light.

The *Electron Gyromagnetic Ratio* for its own is a measure of the rotational velocity of the free electron, indeed a spinning object, or in proper designation its spin angular momentum. The defining formula for the Gyromagnetic Ratio of a classical rotating body is:

$$\gamma = \frac{q}{2m}$$

where q is its charge and m its mass. Note that the constant known as *Electron Charge-to-Mass Ratio* is another ratio figure quite close in value to the *Electron Gyromagnetic Ratio*, with its official value set at $0.1758812012 \cdot 10^{10}$ C/ Kg. Although the units are different, the two figures are very similar to one another given that the Electron Gyromagnetic Ratio is equal to $e/2m$ and the Electron Charge-to-Mass Ratio is equal to e/m . Since the mass is an extremely small value with respect to the charge, the divisions by m or $2m$ end up casting almost the same result. From the Quanto-Geometric analytical perspective however, the relevant figure is the Electron Gyromagnetic Ratio and that is the one we will shortly devote our attention to.

After taking stock of the above facts, one is left to wonder why these very specific quantities have something to do in particular with the electron to start with. Why do these quantities not qualify as universal physical constants for U₃ just as the *Planck constant* and the *Stephan-Boltzmann constant*? The truth is that this is exactly the case, because the electron is the smallest fermionic particle (mass dominant particle) and because the natural physical quantities associated with it potentially become base units for the fermionic world. So therefore the physical quantities associated with the *h quantum* represents the natural units for the mass-less (space dominant) world of U₃ objects whereas the quantities associated with the electron represent the natural units for point-like or mass dominant objects. One may justifiably view the latter equally as a form of quantization, as it in fact is.

By giving primacy to the units of our usual measurements of physical quantities, instead of the cardinality of the quantities themselves, most physicists may have never noticed that the Plank constant is the inverse of the Electron Compton Wavelength and that the Stephan-Boltzmann constant is the inverse of the Electron Gyromagnetic Ratio. The premise that we have set forth corroborates this result in many ways. Suffice it to reexamine the relations *(5-1)* and *(5-2)* for that determination. Effectively if we normalize those two relations by setting the Potential Energy Density E_p equal to 1 and the Kinetic Energy Density equal to 1 (their being equal is what quadratic equalization means), then:

$$h = 1/v \quad \text{and} \quad \sigma^* = 1/T$$

where we assimilate v to the Electron Compton wave function and T to the Electron Gyromagnetic Ratio. While the first assignment may be readily understood, the second may seem at first objectionable. Until we realize that Temperature is defined as the average kinetic energy of a population of objects and that the spin component of the motion of these objects accounts for the most part for that average quantity. While both translational or linear momentum and angular momentum of individual elements in a population of free particles may hypothetically be reduced to a minimum or null, their gyromagnetic moment (spin) is a non-reducible, establishing thereby the basis for kinetic energy density or T as a necessary physical quantity or a characteristic of matter of great consideration (even when we are considering a single object). We shall note, by the same token, that while great emphasis is usually placed on the energy distribution of a radiation spectrum because of its historically revolutionary h *quantization* implications, the role of thermodynamic temperature and the shift of the whole spectrum as the thermodynamic temperature increases in accordance with the Stephan-Boltzmann law are usually understated. All four of these constants that we come to witness thru pure radiation spectra denoted black body radiation are equally important and remit to two distinct bases of quantization of matter as normed by the Quanto-Geometric Hyper Quadratic Operator.

Unlike what has become customary in academia, it is important when representing these spectra to not only present the various energy distribution curves but also the thermodynamic temperature variation curve going thru the peaks of these individual spectra as the second significant tenet of these experimental artifacts (Fig. 5.8).

What's more, these two orders of phenomena come again to be manifest in the value of the spin of both the electron and the quantum of action (the minimal photon). It turns out that the expected value for the spin angular momentum from a "classical" approach is quite different from the experimental value, which was found to necessitate a g correction factor and most importantly was found to be a function of $\hbar$. Interestingly enough, the reduced Planck constant figure is in itself a gyric measure *(h/2π)*, and the quantum mechanic measure for the electron's spin angular momentum is expressed as:

$$p_s = \sqrt{s(s+1)} \cdot \hbar$$

where s is the spin quantum number. The formula makes $\hbar$ a natural scale for that spin. The only value that comes to satisfy experimental results is $s = \frac{1}{2} \hbar$. Now we recall that the norm of the Hyper Quadratic Operator in developing the work of creation at this stage was the selection of two s values of the Quanto-Geometric Function, one of which was $\frac{1}{2} \sigma$. We hold right there the reason for the determination of the spin of the electron at that particular value in Quantum Mechanics. If this is true, the counterpart object, the quantum of action or photon particle, should also have a spin and the value of that spin should be 1 (or *1 σ* or simply *σ*). Well, the photon particle is known to have a full spin value of *1ħ*, in an unmistakable demonstration that those two points of the real image map of our generating work function are the origin of spinors or the spinor space as known in Quantum Mechanics.

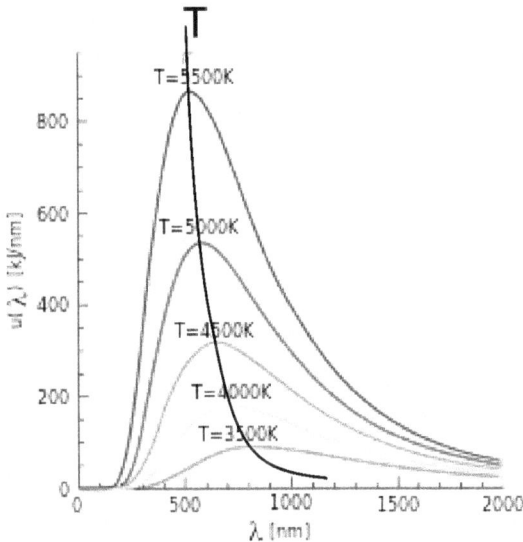

Fig. 5.8 Black body radiation spectrum showing the Temperature function

The above parallel does not necessarily mean that $\hbar$ is equivalent to the standard deviation parameter σ of the Quanto-Geometric Mappa Mundi Function, just as the classical Gyromagnetic Ratio is not exactly interchangeable with the same figure in Quantum Mechanics.

5.10. The Hyper Space Operator and the Interaction Coupling Constants

Within the work of creation, the Tri-Valent Hyper Space Operator is entrusted with the development of space shelling or domain dimensioning in U_3. In its capacity of a space-dominant entity it is best positioned to undertake the task of provisioning spatial encapsulation to the Universe in development. As a reminder the norm of this Operator is *space overture* or *space de-ployment (un-folding)*. This Operator correspondingly takes the following four actions, all consistent with its norm and the nature of the object under operation:

1. It instantiates an image of the Mappa Mundi function, which is appealed to play the role of a template.

2. It combines the Mappa Mundi image's derivative function with a parabolic function in order to generate the final map of construction of U_3. The choice of a parabolic conic function over the other conic functions is geared by the 3-dimensional nature of the Universe in construction *(n = 3)*, which a figure of symmetry base 3 such as the parabola will provide the best generating matrix for.

3. By prescribing the selection of the parabolic function of construction from within the conic sub-functions embedded in the Mappa Mundi function, it poses thereby a requirement for that function to be a Quanto-Geometrically normalized parabola.

Ultimately the end operation that accomplishes the generation of U_3 is given in a Transform T_9 such that:

$$T_9 \to \frac{d[Q(s)]}{ds} \bullet y_p$$

We are going to handle that transform as an $M_3(s)$ function because the parabolic function is ultimately reduced to a simple coefficient in the course of the operation, as expressed below:

$$M_3(s) = \frac{dQ(s)}{ds} \bullet k_p$$

We will shortly see the reason for k_p, a parabolic coefficient, instead of the full parabolic function y_p. In general, the resulting individual transformations create 8 *shells* of matter across U_3, namely the Particle shell, the Atom Shell, the Molecular Shell, The Crystal Shell, the Karyot Shell, the Astral Shell, the Star-System or Solar Shell and the Galactic Shell. We might add the Quark Proto-Shell to make justice to it, notwithstanding its limited shell development. What these shells specifically stand for are whole domains of *de-ployed* space or capsules of matter which are embedded within one another. Below is the hierarchical organization they present in our physical universe:

(Shell 0 → Quarks)

Shell 1 → Particles

Shell 2 → Atoms

Shell 3 → Molecules

Shell 4 → Crystals (Materials)

Shell 5 → Karyots

Shell 6 → Astral Bodies

Shell 7 → Star Systems

Shell 8 → Galaxies

The assignment map for the Transform T_9 is:

$(s_0 = 0 \rightarrow$ Shell 0 – Quarks)

$s_1 = \pm \sigma/2 \rightarrow$ Shell 1 - Particles

$s_2 = \pm 2\sigma/7^{1/2} \rightarrow$ Shell 2 - Atoms

$s_3 = \pm \sigma \rightarrow$ Shell 3 - Molecules

$s_4 = \pm \sigma(8/5)^{1/2} \rightarrow$ Shell 4 - Crystals (materials) or Lattices

$s_5 = \pm \sigma(5/2)^{1/2} \rightarrow$ Shell 5 - Karyots

$s_6 = \pm 2\sigma \rightarrow$ Shell 6 - Astral Elements

$s_7 = \pm \sigma 7^{1/2} \rightarrow$ Shell 7 - Star Systems

$s_8 = \pm 4\sigma \rightarrow$ Shell 8 Galaxies

The match per our interval definition for the Quanto-Geometric Function is:

(Shell 0 $\rightarrow$]0, $\sigma/2$[ )

Shell 1 $\rightarrow$ [$\sigma/2$, $2\sigma/7^{1/2}$[

Shell 2 $\rightarrow$ [$2\sigma/7^{1/2}$, σ[

Shell 3 $\rightarrow$ [σ, $\sigma(8/5)^{1/2}$[

Shell 4 $\rightarrow$ [$\sigma(8/5)^{1/2}$, $\sigma(5/2)^{1/2}$[

Shell 5 $\rightarrow$ [$\sigma(5/2)^{1/2}$, 2σ[

Shell 6 $\rightarrow$ [2σ, $\sigma 7^{1/2}$[

Shell 7 $\rightarrow$ [$\sigma 7^{1/2}$, 4σ[

Shell 8 $\rightarrow$ [4σ, ∞[

We may notice in the above list of shells that *quarks* are annotated pertaining to *Shell 0*. However they do not matter-of-factly form a shell. Shell 0 is not a shell proper per determination of the assigned interval which is open on both ends: it is a *proto-shell*. There is no finite starting point for an *s-interval* for this would-be domain, s equal to 0 being undefined or forbidden by the model. Hence no possible finite $dQ(s)/ds$ value at that point which could serve as a constant. Effectively the first value of *s* within each interval is what is going to yield a physical constant, a quantity incarnated by the corresponding value of the $M_3(s)$ function. Because quarks do not or cannot have a defining physical constant, hence a space domain of their own, they come to behave as *glued* to one another and *confined* within the internal or endogenous boundaries of the next space *de-ploy*, the

Particle Shell. The so-called phenomenon of *confinement* means that they can never be manifest in free space of their own but only as sub-components of Shell 1 elementary particles. They may only have a life as indivisible parts of particles in what construes as the closest representation of pure scalar elements in U_3, if ever possible. Their inner space value is close to 0, albeit not 0. This circumstance does not preclude them from having behavior however, specifically Quanto-Geometric behavior. They remain thoroughly subject to Quanto-Geometric spectral organization based on the 9 typified primitives and other determinants of the model as we shall see in Volume II. This is by all accounts the origin of the phenomenon of *quark confinement*, empirically observed and subsequently posited, but not theoretically derived, in the Standard Model of particle physics.

As to *Quasars*, which are not explicitly mentioned in the above presentation, they are part of the Galaxy class since they are known to have galactic identity as proto-galaxies.

Does this entire construction all amount to just the author's view, a speculation or a mere vanity? Let us examine the creation work function $M_3(s)$ for an answer to that question.

5.11. The Parabolic Coefficient as a Generator

From previous developments, we know that the Quanto-Geometric $Q(s)$ function encloses a parabola among its embedded conic sections. We have stated right above that the Tri-Valent Hyper Space Operator has elected to incorporate this parabolic function within the transformation at work to give birth to U_3. On a Cartesian coordinate system, the expression of this embedded parabola would be:

$$y = \frac{1}{\sigma\sqrt{2\pi}} + \frac{s^2}{4p} \text{ , with } \frac{1}{\sigma\sqrt{2\pi}} \text{ being the y intercept.}$$

We know however that the Quanto-Geometric $Q(s)$ function is not defined for $s = 0$, so that no y intercept exists for the embedded parabola. On the Quanto-Geometric coordinate system the expression of this embedded parabola centered on the q axis reads:

$$y = \frac{s^2}{4p}$$

The *latus rectum* point of the parabola also belongs to $Q(s)$ as s_3. We remind the reader that the latus rectum is the cord that goes thru the focal point of the parabola and stands orthogonal to the y axis. Its distance to the vertex of the parabola (a point of discontinuity in our function) is p. It can be demonstrated that the length of the latus rectum at each side of the y-axis is $2p$, for a total length of 4p for that special cord. Since the point $s = 2p$ of the parabola also belongs to the $Q(s)$ as s_3, it follows that:

$$s_3 = 2p \text{ and from } Q(s) \ s_3 = \frac{2\sigma}{\sqrt{7}}.$$

So that: $2p = \dfrac{2\sigma}{\sqrt{7}}$ or $p = \dfrac{\sigma}{\sqrt{7}}$.

Replacing p by its value in the equation of normalized parabola yields:

$$y = \frac{s^2}{4p} = \frac{s^2}{4} \cdot \frac{\sqrt{7}}{\sigma}$$

At s₃, $y_p = \dfrac{\sqrt{7}}{4\sigma} s^2 = \dfrac{0.66114378278}{\sigma} \cdot s^2$ or:

$$y_p = \frac{0.66114378278}{\sigma} \cdot s^2$$

The eventual graph of y_p follows in Fig. 5.9.

Now, it is important to note that σ does not have the same dimension as s in the above expression of the parabola. The σ parameter is the value of s at inflexion point in the $Q(s)$ function, which lays much beyond any shared point with the embedded conic sections. Therefore the point $s = \sigma$ does not and cannot belong to the normalized parabola. Thus although σ is an s value, it cannot stand as such in the expression of the parabolic function. As a result within that expression, the entire term $0.66114378278/\sigma$ which appears as a coefficient of s^2 can only have the value or weight of a coefficient as a "dimensionless" scalar. We denote that term as follows:

$$k_p = \frac{0.66114378278}{\sigma}$$

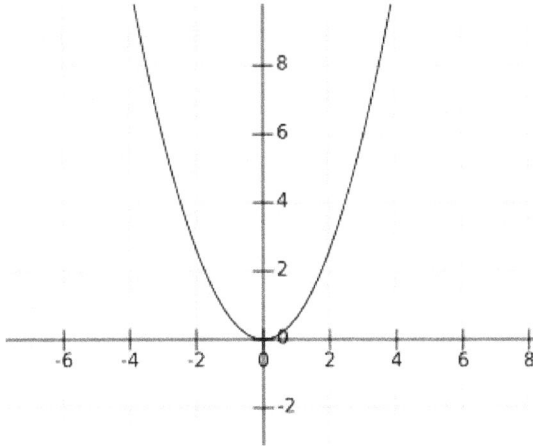

Fig. 5.9 Graph of y_p

The parabolic coefficient k_p is instead going to play that eminent role in the generation of U_3 as we shall see shortly, in substitution of y_p to which the role is *naturally* reserved.

5.12. Derivative of the Grand Mappa Mundi Function

As previously noted, the work of Shell de-ployments in U_3 is performed according to the end function $M_3(s)$:

$$M_3(s) = \frac{dQ(s)}{ds} \bullet k_p$$

The first-order derivative expression of $Q(s)$ reads as follows:

$$\frac{dQ(s)}{ds} = \frac{1}{\sigma\sqrt{2\pi}} \cdot \frac{-2s}{2\sigma^2} \cdot e^{\frac{-s^2}{2\sigma^2}}$$

$$\frac{dQ(s)}{ds} = \frac{-s}{\sigma^3\sqrt{2\pi}} \cdot e^{\frac{-s^2}{2\sigma^2}}$$

$M_3(s)$ becomes after due replacements:

$$M_3(s) = \frac{-s}{\sigma^3\sqrt{2\pi}} \cdot e^{\frac{-s^2}{2\sigma^2}} \bullet \frac{0.66114378278}{\sigma}$$

$$M_3(s) = 0.66114378278 \bullet \frac{-s}{\sigma^4\sqrt{2\pi}} \cdot e^{\frac{-s^2}{2\sigma^2}}$$

See the graph in Fig. 5.11 for a pictorial representation of the $M_3(s)$ function. We may now replace s successively by the critical values defining the intervals in order to obtain the defining constants for each physical Shell of U_3 as their building blocks. We shall point out that what we call the defining values of these intervals are the first or initial value of each. The Hyper Space Operator assigns each of these initial values to the generation of a Shell, thereby creating the defining physical constant for each Shell thru the Norm of the Operator, which amounts here to simply the value of the $M_3(s)$ function. Under these circumstances s creates the space *de-ploy* or composition and $M_3(s)$ establishes the pillar of the Shell edifice, the physical constant.

5.13. Shells of U_3 and Their Physical Constants

We are going to compute the value of $M_3(s)$ for each interval s_i corresponding to the different physical Shells making up U_3.

At s_1, Shell of Particles or S_1:

$$s = \frac{\sigma}{2}, \quad M_1(s_1) = 0.66114378278 \bullet \frac{-s}{\sigma^4\sqrt{2\pi}} \cdot e^{\frac{-s^2}{2\sigma^2}}$$

$$= -\frac{0.66114378278 \cdot 0.1760326634}{\sigma^3}$$

$$= -\frac{0.1164346625}{\sigma^3}$$

At s_2, Shell of Atoms or S_2:

$$s = \frac{2\sigma}{\sqrt{7}}, \ M_2(s_2) = 0.66114378278 \bullet \frac{-s}{\sigma^4 \sqrt{2\pi}} \cdot e^{\frac{-s^2}{2\sigma^2}}$$

$$= -\frac{0.66114378278 \cdot 0.2266245234}{\sigma^3}$$

$$= -\frac{0.1498980325}{\sigma^3}$$

At s_3: Shell of Molecules or S_3:

$$s = \sigma, \ M_3(s_3) = 0.66114378278 \bullet \frac{-s}{\sigma^4 \sqrt{2\pi}} \cdot e^{\frac{-s^2}{2\sigma^2}}$$

$$= -\frac{0.66114378278 \cdot 0.2419707245}{\sigma^3}$$

$$= \frac{0.1600485904}{\sigma^3}$$

At s_4: Shell of Crystals or S_4:

$$s = \sigma, \ M_4(s_4) = 0.66114378278 \bullet \frac{-s}{\sigma^4 \sqrt{2\pi}} \cdot e^{\frac{-s^2}{2\sigma^2}}$$

$$= -\frac{0.66114378278 \cdot 0.2419707245}{\sigma^3}$$

$$= -\frac{0.244197976}{\sigma^3}$$

Since for the Shells beyond S_4, no physical constant is known, although they very much exist, we will stop the list with this Shell. An example of this computation in full is shown below for illustrative purposes; it is based on the S_2 Shell of Atoms.

$$M_{2(s_2)} = M_2(s_2) = 0.66114378278 \cdot \frac{-s}{\sigma^4 \sqrt{2\pi}} \cdot e^{\frac{-s^2}{2\sigma^2}}$$

$$= 0.66114378278 \cdot \frac{-(2\sigma/\sqrt{7}) \cdot e^{\frac{-(2\sigma/\sqrt{7})^2}{2\sigma^2}}}{\sigma^4 \sqrt{2\pi}}$$

$$= 0.66114378278 \cdot \frac{-(2\sigma/\sqrt{7}) \cdot e^{\frac{-(4\sigma^2/7)}{2\sigma^2}}}{\sigma^4 \sqrt{2\pi}}$$

$$= 0.66114378278 \cdot \frac{-(2\sigma/\sqrt{7}) \cdot e^{-(2/7)}}{\sigma^4 \sqrt{2\pi}}$$

$$= 0.66114378278 \cdot \frac{-(2\sigma/\sqrt{7}) \cdot e^{-(2/7)}}{\sigma^4 \sqrt{2\pi}}$$

$$= 0.66114378278 \cdot \frac{-2\sigma \cdot e^{-(2/7)}}{\sigma^4 \sqrt{7}\sqrt{2\pi}}$$

$$= 0.66114378278 \cdot \frac{-0.2266245234}{\sigma^3}$$

$$= -\frac{0.1498980325}{\sigma^3}$$

	3-D Universe → s = $2/7^{1/2}$ σ
$Q(s) \rightarrow s_3$	0.299796065/ σ — Speed of Light

	Spinor space → $s_1 = \frac{1}{2}$ σ	Spinor space → $s_1 = 1$ σ
dq/ds	0.1760 3266/ σ² — Electron Gyromagnetic Moment	0.2419707245/ σ² — Electron Compton Wavelength
$(dq/ds)^{-1}$	5.680763* σ² — Stefan-Boltzman constant	0.413273135* σ² — Plank constant (in eV s)

	Elementary Particles - Shell 1	Atoms - Shell 2	Molecules - Shell 3	Crystals - Shell 4
$k_{\sigma}^{*} \cdot (dq/ds)$	0.116643466/ σ³ — Fine Structure constant	0.1498980 325/ σ³ — Inverse of Gravitational constant	1.600485904/ σ³ — Elementary Charge	0.2441197/ σ³ — Van Der Vaal constant

	Karyots - Shell 5	Astral Elements - Shell 6	Star Systems - Shell 7	Galaxies - Shell 8
$k_{\rho}^{*} \cdot (dq/ds)$	0.2988415566/ σ³ — unknown	0.2856933408/ σ³ — unknown	0.14757590 /σ³ — unknown	0.00566530 / σ³ — unknown

	Elementary Particles - Duplet	Atoms - Duplet	Molecules - Duplet	Crystals - Duplet
$(k_{\rho}^{*} \cdot dq/ds)^2$	0.1355703/ σ⁵ — Rydberg Energy unit R_{∞}	0.02246942014/ σ⁵ — Molar volume of Ideal Gas	0.02561551/ σ⁵ — unknown	0.059594428/ σ⁵ — unknown

Table 5.3 Table of physical constants deriving from Quanto-Geometric functional constructs. We have muted the negative sign of the quantities, whose principal meaning concerns the direction of correlational displacements.

Reported in Table 5.3 are all these values per Shell 1 up to Shell 4; it includes as well other related figures which contribute to the makeup of all physical constants intervening in the composition of the U_3 universe. Effectively the derivative function itself produces physical constants so much as its inverse and the inverse of the generating function $M_3(s)$ do. This table allows us to appreciate the very important fact that the known physical constants are not random values only computationally related to one another but obey to an order of hierarchies which are the vertical and horizontal Quanto-Geometric hierarchies set forth in this work. Some of these figures may be readily recognizable by the reader but others may not. We will next address each in more details.

We shall also note that these figures are harmonized due to their common origin, being all expressed in parametric form as a composition between a scalar and σ the standard deviation of the Grand Mappa Mundi Function. Consequently they will not be in an absolutely exact match with the known figures but they will match by an acceptable order of "pure magnitude", or to be more exact as **figurate** numbers. The most striking fact remains how from a single and relatively simple mathematical-physics system all these constants are naturally derived, raising the question of which side needs a correction if need be as to the orders of magnitude.

The reader will find in Appendix I a list of all known physical constants published by the U.S. National Institute of Science and Technology. It is a rather furnished list, however one shall not lose sight of the fact that there is only a handful of foundational physical constants, the most preeminent of which is the speed of light.

5.14. Derivation of the Newton Gravitational Constant

The official value of the gravitational constant G is $6.6720 \ 10^{-11}$ N-m^2/kg^2, which makes G^{-1}, the inverse of G, equal to: **$0.1498800959 \ 10^{-12}$ kg^2/N-m^2**.

Our table reports G^{-1} as a Shell 2 constant with the value of: **$0.1498980325/\sigma^3$**. Now, one can see that the two values are appreciably close to one another except for the dimensions they are expressed in. The dimension N-m^2/kg^2, with the inverse kg^2/N-m^2, is a view of G. That is how we historically came to know of it by physical measurements based on our understanding of physical nature.

The most significant flaw of that view consists in that it makes no distinction between the qualities of space in each space shell, nor does it understand the nature of space as dictated by the vertical hierarchy. It is a rather significant fact that G represents a physical constant in operation in Shell 2, the shell of Atoms. It means that contrarily to what is expected by both the Standard Model and the Quantum Gravity models, gravity does not exist at the particle level. No such thing there is as gravity being so weak at the particle level that it becomes lost in the presence of stronger forces as the current belief holds. *There is simply NO gravity at the particle level.* If we expect gravity to be detectable as gravitational waves similar to the wave function of particles, and that gravity should be in effect at the level of interaction between free particles, then it is a lost quest. On the other hand, if we restrict the statement strictly to the Atom shell as the lower limit of exercise of gravity, or if we look for gravity within the interaction taking place between particles *in situ* inside the atomic edifice (not free particles), then we are off to a good start.

Finally the fact that we commonly know of and find use for G and not G^{-1} is equally telling. It reveals that the view which leads us to G is at the very least a backward view from a rear vantage point, not to say that it is a skewed view. Since we are looking at everything occurring in sub-shells of space from another higher shell without making the necessary preventions, exemptions and distinctions, it is no wonder that the resulting statements or conclusions are somewhat falsified, distorted or otherwise inadequate.

5.15. Derivation of the Electric Charge of Electrons

The official value of the electron charge constant e currently is **1.60217656 10^{-19}** Coulomb. As per our table e is a Shell 3 constant with the value figure of: **1.600485904/σ^3**.

Remarkably, it is not a constant emanating from Shell 1, the Particle Shell, as one might have expected. It is a building block for the Molecular shell. Be that as it may, as soon as we remember that the electrostatic systematic between atoms' electronic shells is at the heart of molecular elements then the surprise element might fade away. This all means that what is happening in the Molecular Shell at its formation is a condition whereby the interaction between atoms is pre-set by an electrical systematic, the smallest unit or form of which is the unit-Coulomb. In other words the e constant is a building block for the Molecular Shell, and not a quality inherent to the *electron*. That is the very reason why the proton also demonstrates electrical charge at exactly the same value.

The property of charge is a property of the molecular medium not of the particles residing therein. We recognize that this conclusion will be a difficult fact for physicists to accept, given that so much has been built on the concept of charge being inherent to elementary particles themselves. However few have historically asked themselves the question: how come the value of the electrical charge of an electron is exactly the same as that of a proton which is a particle with much larger mass? What in the particle makes a charge unit? Is the charge an artifact of mass, is it an artifact of the particle's wave function? Where does it come from?

The nature of space within the Molecular medium is the origin of this constant. It is the driver that orchestrates the systematic between atoms at formation of Molecules and their subsequent behavior during their life-cycle. For those who will still resist the idea of charge not being inherent to particles, we invite them to find a reason or an element capable of explaining the property or quality of their alleged charge, the origin of that charge, not just rely on experimental measurement of quantities, which are designs from our own constructs. Some other polemic facts with no answer are well known about particles in relation to their electric charges or lack thereof. We will defer this analysis for treatment in Volume II.

5.16. Derivation of the Fine-Structure Constant and the Fermi Coupling Constant

Interestingly enough, these two constants are very closely related to one another when viewed thru the prism of Quanto-Geometric treatment. For one, they are both coupling constants, and secondly their value is almost identical when estimated or computed suitably. The Fermi coupling constant is a mediator for the electro-weak force (radioactive decay) whereas the fine structure-constant, also known as the electro-magnetic coupling constant, is said to be a mediator for electro-magnetic interaction between elementary charged particles.

5.16.1 The Fine-Structure Constant

The official value of the fine-structure constant α currently is **7.297352568 10^{-3}**. This may be the only wholly computed constant, after the permittivity and permeability of space, both of which are incidentally part of the formula for α, as reported below.

The formula is:

$$\alpha = \frac{e^2 / \hbar c}{4\pi\varepsilon_0},$$

where e is the elementary charge, $\hbar$ is the reduced Plank constant. Since c the speed of light and ε_0 the permittivity of space are formulated as:

$$c = \frac{1}{\sqrt{\mu_0 \varepsilon_0}}$$

and the reduced Plank constant is valued at $\hbar = h/2\pi$, the formula for α becomes after these replacements:

$$\alpha = \frac{\mu_0 c e^2}{2h}$$

By replacing these constants with their known values in the α formula above, α comes out to be:

$$\alpha = \frac{(1.602176565 \cdot 10^{-19})^2 \times 299792458 \times 4\pi \cdot 10^{-7}}{2 \times 4.135667516 \cdot 10^{-15}}$$

$$= \frac{2.566969745 \cdot 10^{-38} \times 376.7303135}{8.271335035 \cdot 10^{-15}}$$

$$= \frac{9.670553169 \cdot 10^{-36}}{8.271335035 \cdot 10^{-15}}$$

$$\alpha = 0.116916654 \; 10^{-20}$$

The Plank constant h is generally chosen to be **6.6256 10^{-34} J-s** when computing the value of α. We have chosen instead in computing the formula its value expressed in eV-s (electron-volt–second), a more standardized or primary dimension than the J-s (Joule-second) dimension, because it is directly related to e, the electron or its charge. As shown on Table 5.3, the Quanto-Geometrically derived value for the constant is: **0.1164346625/σ^3**.

While the value of the fine structure constant can be fully computed from any of the de-fining formulae, quantum electrodynamics has provided ways to measure this constant, principally via the quantum Hall effect. As usual with measurements of physical con-stants, they have historically gone from appreciably inaccurate to appreciably accurate as the technological means undergo refinement. It is fair to say that this constant is one that has most fascinated physicists, even to the point of bewilderment, different visions of what it exactly means having been debated. If physicists are loath to numerology, re-garded by them as a suspicious pseudo science, the fine structure constant had them en-gage openly in that exercise. Here is what Richard Feynman, one of the pioneers of Quantum Electrodynamics, wrote on this subject in 1985:

"There is a most profound and beautiful question associated with the observed coupl-ing constant, e – the amplitude for a real electron to emit or absorb a real photon. It is a simple number that has been experimentally determined to be close to 0.08542455. (My physicist friends won't recognize this number, because they like to remember it as the inverse of its square: about 137.03597 with an uncertainty of about 2 in the last decimal place. It has been a mystery ever since it was discovered more than fifty years ago, and all good theoretical physicists put this number up on their wall and worry about it.) Im-mediately you would like to know where this number for a coupling comes from: is it re-

lated to pi or perhaps to the base of natural logarithms? Nobody knows. It's one of the greatest damn mysteries of physics: a magic number that comes to us with no understanding by man. You might say the "hand of God" wrote that number, and "we don't know how He pushed his pencil." We know what kind of a dance to do experimentally to measure this number very accurately, but we don't know what kind of dance to do on the computer to make this number come out, without putting it in secretly!"

One may note with a smile that the accuracy of the number today puts it at 0.07297352568 as previously mentioned, not his "very accurate" 0.08542455 commonly accepted at the time! But his remarks are admittedly otherwise quite telling. This quantity so intrigued physicists, not just theoretical physicists, that Wolfgang Pauli, known by and large for his contribution to Quantum Mechanics, took it to the point of collaborating with Carl Gustav Jung, a towering contemporaneous figure in the field of Psychology, in order to unravel the transcendental meaning of this quantity.

Because this constant lends itself to so many different deriving formulae implicating thereby different genericity, its physical nature has come to be viewed in many different ways by physicists. In some interpretations, it is considered to typify interaction between free particles such as free electrons and photons. When derived from the application of classical methods, it is viewed as a characterization of a *massless* charge on a circular track embedded in its own radiation field and expressing the strength of the electro-dynamic interaction. Quantum Mechanics has it that it characterizes the interaction between a charge and its field, a form of kinematic that replaces the punctual charge by a charge distribution rather which leads to oscillatory behavior. It is a rather remarkable fact that this constant can be valid without any correction both in Classical energy physics and in Quantum Mechanics, knowing the conceptual steep jump laying between the two fields. More on this later. Its application in physics is extremely widespread, perhaps second only to c, the speed of light, or h, the Planck constant. The truth to this matter is that the fine-structure constant is a characterization of the quality of space intervening in the buildout of not only electrons and photons, but all 3-dimensional particles as pertaining to Shell1 of U_3, the Quanto-Geometric Particle Shell. If we were to be more precise, we would make the statement that it corresponds to the specific value of the so-called $M_3(s)$ composite Quanto-Geometric function at a specific value (s_3) of s.

We do not have to engage in an exercise of numerology about it, having established the Quanto-Geometric number theory that systematically leads us to its province. So the late Richard Feynman, Wolfgang Pauli and all other theoretical physicists with this interest to date are hereby well served! We shall conclude these remarks by stating that the *pencil of God* that R. Feynman referred to and *the way that He pushed it* are indeed the generating Transform materializing the work of Shell generation in U_3:

$$M_3(s) = \frac{dQ(s)}{ds} \bullet k_p$$

5.16.2 The Fermi Coupling Constant

In the Standard Model of particle physics, interaction between sub-particle components that make up particles, which becomes observable in decay or radioactive events where they occur, is developed thru the mediation of a coupling constant. The primary coupling constant for such interaction is the Fermi coupling constant G_F, after the name of its proponent, Enrico Fermi, and proposed since the early days of particle physics. The Electroweak Theory of the Standard Model so restricts the domain of this constant to unstable particles while it constitutes the province of interpretation for all particle decay phenomena, whether they occur in free particles such as the naturally unstable neutron or within the context of a higher edifice such as atomic nuclei.

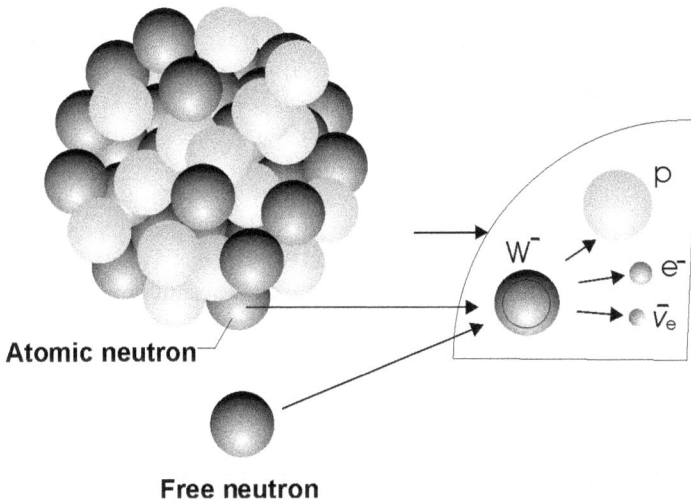

Fig. 5.10 Beta decay of a neutron particle

The formula and value of G_F are as follows:

$$\frac{G_F}{(\hbar c)^3} = \frac{\sqrt{2}}{8} \frac{g^2}{m_W^2} = 0.116637 \times 10^{-4} \, GeV^{-2}$$

where g is the coupling constant (*g-factor*) of the so-called electroweak interaction and m_W the mass of the W boson that mediates the decay interaction. The whole term

$$\frac{G_F}{(\hbar c)^3}$$

is the quantity in use by preference in modern physics, not just the original Fermi's G_F.

Is it any surprise that this value is in a match with the fine-structure constant provided the latter is suitably computed, and in a match as well with the Quanto-Geometrically derived value? Beta decay in an atomic nucleus, resulting from the breakdown of a neutron into several other sub-particles, was the original focus of Enrico Fermi (Fig. 5.10). His proposal, originally a theoretical treatment, was initially rejected, but later gained credit when it became experimentally ratified and other forms of particle decay, aside from neutron decay, could be accounted for by the constant. This and other Fermi's findings were instrumental in paving the way for the development of particle physics. Finally this constant registers as a paramount entity in the composition of all Shell1 elementary particles, whether they are stable or unstable.

5.17. Unifying the Electromagnetic Interaction with the Electroweak Interaction

The many distinct views that physicists have historically held about the identity of the electromagnetic coupling constant have not prevented them from sharing a sense of its possible relationship with the electroweak force. Reason why a so-called *unifying electroweak theory* has been developed in order to establish a single scale for both types of interaction, the difference between one and the other being a simple question of level of energy harbored by each. Effectively in the unified electroweak theory the electromagnetic interaction is part of a mixture of interactions associated with electroweak fields so that the strength of the electromagnetic interaction becomes a function of the strength of the energy field. The observed value of α relates therein to the energy level of the mass of the electron. Since the electron is the lightest charged object, then α is ranked at ground level or energy zero in the scale, leaving the Fermi coupling/energy constant higher up in the scale.

This construction has been plagued with shortcomings because the expected projections do not realize in the actual behavior of matter, forcing physicists to engage in a series of ill-fated extrapolations called *Renormalization, Regularization* and other misfit engenders. These developments drew sharp criticism from some of the early and renown formulators of Quantum Electrodynamics. Irate Paul Dirac explains as late as 1975:

*Most physicists are very satisfied with the situation. They say: 'Quantum electrodynamics is a good theory and we do not have to worry about it anymore'. I must say that I am very dissatisfied with the situation, because this so-called 'good theory' does involve neglecting infinities which appear in its equations, neglecting them in an **arbitrary** way. This is just **not sensible** mathematics. Sensible mathematics involves neglecting a quantity when it is small – not neglecting it just because it is infinitely great **and you do not want it**!*

Richard Feynman, another lead physicist in the formulation of the theory, further added:

*The shell game that we play ... is technically called 'renormalization'. But no matter how clever the word, it is still what I would call a dippy process! Having to resort to such hocus-pocus has prevented us from proving that the theory of quantum electrodynamics is mathematically self-consistent. It's surprising that the theory still hasn't been proved self-consistent one way or the other by now; I suspect that renormalization is **not mathematically legitimate**.*

Finally, Lewis Ryder had this to say about the state of affairs that lasted to this day:

*In the Quantum Theory, these [classical] divergences do not disappear; on the contrary, they appear to get worse. And despite the comparative success of renormalization theory the feeling remains that there ought to be a **more satisfactory way** of doing things.*

There are many reasons to the failure of Quantum Electrodynamics to unify the "electroweak forces", let alone the adjunction of the other two fundamental forces in the act. The primary reason is that its scale of interaction failed to recognize the extent to which space vacuum behaves as a physical variable, and secondly that the energies are distributed across distinct real Shells of space that are completely integral to the composition of matter. Lastly it failed to understand that the electric field does not exist across all Shells except for one. I might add that the recognition that the authentic value of α is not 1/137 but **0.116916** which puts it not only close to but in superposition with the figurate **0.116637** of the Fermi coupling constant would have been a great start! Finally, let it be resoundingly proclaimed, there is a high price to pay for abandoning the key signature that has made Quantum Mechanics otherwise so successful: the uncompromising and unmitigated ban of *time* as a physical variable. To that account one shall add the deep underestimation that Quantum Mechanics for its own ought to imperatively and permanently subscribe to that school of thoughts and adopt its implications as a matter of necessity.

5.18. The Fine-Structure Constant and the g-factor Cluster

In particle physics, one important defining quantity for an elementary particle (leptons and bosons) is its *magnetic moment*, also called *magnetic dipole moment*. This notion is tightly related to that of fine-structure constant (α), otherwise called electromagnetic coupling force, which we have just studied above. The clearest view of what the fine-structure constant represents is that it stands for the originator of the internal binding force operating inside of particles, which the Standard Model of particle physics believes to be of magnetic nature. As such it is the originator of all forms of observable magnetic moments known to particles. The spin of the particle is a contributor to its magnetic moment; for an atomically bound particle, the angular momentum is another contributor, and the composition between spin and angular momentum represents a third contributor in that case. The spin or gyromagnetic moment is the single contributor to the magnetic moment for a free particle and remains the chief contributor to that of an atomically bound particle.

In particle physics, the magnetic moment is generally expressed in terms of a g-factor, which measures the difference between the expected or predicted value and the one actually observed or measured experimentally. That differential quantity is termed *anomalous* magnetic moment. The disparity may very well be due to the wrong quantity being used for alpha as we pointed out earlier. It has been fashionable as a matter of practice to use the quantity **7.297 352 5698 10^{-3}** or its inverse in physical calculations. We have shown however that the authentic value of alpha is **0.1164346625/ σ^3**. That stands for the very reason why, after applying the respective g-factor correction, all the experimentally measured magnetic momenta of elementary particles narrowly approach the Quanto-Geometrically derived value as reported in Table 4.4 below.

Constant	Value
Fermi coupling constant	0.1166 364 e-4 GeV^-2
Electron magnetic. moment anomaly	0.1159 652 180 76 e-2
Muon magnetic moment anomaly	0.1165 920 91 e-2
Helion magnetic mom. to Bohr magneton ratio	0.1158 740 958 e-2
Shielded Helion magnetic moment to Bohr magneton ratio	0.1158 671 471 e-2
Sackur-Tetrode constant (1 K, 101.325 kPa)	0.1164 8708 e1

Table 5.4   The g-factor cluster for the fine-structure constant

Central to those quantities is the standard value of alpha with which they present figurate matches to different extents. We shall point out as well that even the non-standard value of alpha relates to the Quanto-Geometric figure via division by 2π as computed below:

$$\frac{\alpha}{2\pi} = \frac{7.2973525698 \cdot 10^{-3}}{2\pi} = 0.11614097 \cdot 10^{-2} = a,$$

where a is the *anomalous* moment. Within our framework, the division by 2π relates to the 3-dimensional trigonometric symmetry of the Quanto-Geometric model. The above figure is theoretically attributed to the electron's anomalous magnetic moment in the Standard Model. The experimentally measured value is: 0.115965218076 10^{-2}. For the Muon particle, a heavy electron, the anomalous magnetic moment is: 0.116592091 10^{-2}, which puts it quite close in figurate match to the value of the Fermi coupling constant:

0.1166364 10^{-4}. Incidentally, the measurement of the muon lifetime in neutron degeneracy (radioactive decay) provides the most precise determination of the Fermi coupling constant, the vehicle of the Electroweak force. The W boson, the first intermediate degeneracy product, plays a crucial role in neutron decay just as the W and Z bosons play a significant role in the magnetic moment of the muon particle. We may thereby conclude that the fine-structure constant is a first-order regulator of bosonic behavior (photons and interaction bosons). This suggests that the Unifying Electroweak theory, if it is still to be completed to flawlessness, is better orchestrated along the line of bosonic behavior, with alpha as a prime orchestrator, than along the line of magnetic behavior, which is keenly challenged by the neutron electric paradox.

Should we consider here the g-factor for the neutron and the proton as well? This analysis has already been completed in the review of the Fermi constant. For composite particles such as neutrons and protons, the order of the sub-components of these particles is the stage where the fine-structure constant develops its intervention by scaling the energy of interaction of photons and interaction bosons. One may choose to model the systematics of this interaction in the terms undertaken by the Electroweak Theory. However it is clear that the fine-structure constant is a paramount entity in the internal configuration of all particles, both stable and unstable, since all other quantities that one may construe as a constant in the realm of sub-particles are strictly dependent on alpha as the chief scaling purveyor.

Lastly there is the curious appearance of the Sakur-Tetrode constant in this landscape. The Sakur-Tetrode view is an attempt to frame the state of an ideal classical gas (monoatomic gas) in the extreme condition of temperature as low as 1 K, extremely close to the absolute 0 mark. In this regime the classical realization of a gas (its molecular-Shell spread) is suppressed because the quantum mechanical effects take over. The molecular-level entropy entirely collapses into particle-Shell level entropy and the scalarity of the atom singlets is forced to behave similarly to elementary particle scalarity. This is of course a Quanto-Geometric translation of what is reflected in the Sakur-Tetrode equation, below reported, in terms of the mass of gas particles expressed in atomic units and the presence of the Planck constant in the equation. Those inclusions are equivalent to breaking up the atoms into a sea of small particle-like elements (Shell1 elements) as it were. The very figure of the constant, S/R or entropy over the gas constant, which attempts to capture entropy per unit volume per particle unit, is indeed an expression of particle-level entropy (distribution of entropy per quantum particle). It is worth noting that this constant may be derived as well from pure Quantum Mechanical treatment. It takes its name however after Hugo Martin Tetrode and Otto Sackur who independently derived it in very early 1900's principally from the then prevalent thermodynamical view of physical states. The equation follows:

$$S = kN \ln\left[\left(\frac{V}{N}\right)\left(\frac{U}{N}\right)^{\frac{3}{2}}\right] + \frac{3}{2}kN\left(\frac{5}{3} + \ln\frac{4\pi m}{3h^2}\right)$$

where V is the volume of the gas, U is its internal energy, N the number of gas particles, k the Boltzmann's constant, m is the mass of a gas particle in atomic unit and S the entropy of the gas. At standard condition of pressure *(p= 1 at or 101.325 kPa)*,

$$\frac{S_0}{R} = 0.11648708 \cdot 10^1$$

This figure curiously shows an even wider figurate match with the Quanto-Geometrically derived quantity for the fine-structure constant.

It cannot be overstressed while ending this section that the fine-structure constant α plays a capital role in the internal architecture of elementary particles. In the ignorance of the origin of this number, physicists have understandably arranged to develop the physics of elementary particles on a side track to this capital quantity. By doing so, however, they have run into several impasses requiring them to drop this quantity "secretly" and extemporaneously, in the very word of Richard Feynman, into their equations for those to become sensible and effective. Further, the incorrect valuation of alpha overall has also driven physicists to develop different frameworks for the treatment of radioactivity and the physics of the photon, requiring them to find at the end a difficult unifying umbrella in response to the unmistakable perception that they are forces of the same nature. This clearly transpires in the terminology of *anomalous* magnetic moment attributed to particle's magnetic behavior, whereby the expected value from interpretation does not match the real experimental behavior. In all, one has to be persuaded that when nature does not match the view that one upholds in theory, it is not nature that is *anomalous* but the theory itself.

5.19. Fabric of Space as the Mother of All Motion

Now that we are studying the physical manifestation of the derivative of the Grand Mappa Mundi function, we shall take that opportunity to reinforce the notion of space as the mother of all motion, especially for those who might have had to skip the previous Chapter. We had firmly established the principle that the s variable of the function represents its **independent** variable. In that capacity it is the originator of change. The function or value of $Q(s)$ varies in response to change initiated by the independent variable s. So therefore space itself is auto-dynamic and modeled as such by the independent variable s of the Quanto-Geometric function (the Mappa Mundi).

The derivative of the function as applied to the interpretation of the physical Universe provides a corroboration of that principle. What we have established so far is that the resulting values of the $M_3(s)$ function, which is essentially the derivative of $Q(s)$, for specific values of s, come to be in agreement with the physical constants that govern the development of the physical Universe. Since the derivative of a function is the study of how the function changes or responds to the changes in values experienced by the independent variable, the conclusion then is that indeed s itself is the initiator of change or physical motion. Philosophically the notion of change is at its origin equivalent to or interchangeable with the notion of motion because there *can be no change with no motion,* whether this motion is endogenous or exogenous.

It is easy to see how an exogenous change implies motion since it imperatively implies the passage from one position to another. Less obvious may be the implication of motion in endogenous change since it rather evokes the passage from one state to another. All states, albeit appearing as motionless artifacts, are in fact stationary motion, as all standing waves are. It is because we have not been able to perceive that *state* motion that we have historically assimilated static to absolutely motion-less. Motionless is relative in many ways. Both Fourier analysis and the De Broglie treatment of objects, which associates a wave function to all objects, straightforwardly lead us to that conclusion. Ultimately there is not a single object in absolute rest in the entire Universe, and that alone is a probing fact to our argument.

The problem with the visualization of *nothingness* being able to *move* is a paradigm that defies our capability for high level abstraction, both for the learned and the common man. That motion is not the motion of particle pairs beyond the span of space, as visualized in Quantum Field Theory. It is a capability for innate self-replication which spells auto-generation of motion. Our inability to assert that representation is what explains why we are all still using the baffling abstraction of *time* in our minds. The motion of space is endogenous and not scaled in time. The scale of the motion emanates from the matrix of space itself, so to speak. The inability to sustain that representation is what constitutes the handicap preventing Quantum Electrodynamics and Quantum Field Theory from reaching the goal of unification with General Relativity. By visualizing "fluctuations" in pure vacuum Quantum Field Theory makes an incipient step forward, but attempting to describe the kinematic of a point element in that field with the time variable is making a horrible step backward. We can help ourselves with the assimilation of the paradigm of inner dynamics of space by porting everything we visualize about time to the nothingness of space. It is not going to resolve the issue but it sure is a start. One might otherwise refute our argument by making the following observation: well isn't that what the Laplace transforms do, in shifting the visualization of phenomena in *time* to their visualization in the s domain, the frequency domain? That is a start as well, but just a start, because we are still helping ourselves with the time variable in expressing frequency at the end.

If space can *dilate* as stipulated in Relativity, it certainly can move, because something intrinsic to it has to move in order for it to *dilate*. But then the question is: what is the essence of space motion and what caused that intrinsic motion to stop at end of the dilation process? The truth is that the entire universal space is moving, that is why the speed of light is the same everywhere in U_3: the speed of the perturbation that we call light is indeed the speed of space moving isotropically in all directions. This is not to be confused with expansion or inflationary expansion of universal space in the manner contemplated in Standard Model cosmology. Beyond the question of motion or kinematics, there are equally questions to be raised in the order of structure of the space matrix: i.e. how is the matrix graded and what are the processes that take place inside of the matrix? We believe to have answered an important aspect of these questions with the preeminent symmetry-breaking n sequence and its subsequent segregation of space in intervals of entropy. Further elucidation we will offer as we go. As *unphysical* as this model might appear to common sense, we ought to remain persuaded that it is the one guiding this unraveling of the physical constants now well underway in this study.

5.20. Derivation of the Van Der Vaals *a* Constant

To our consternation, we shall state that, setting aside the very well developed and successful Kinetic Theory of Gases, the study of inter-molecular forces at the root of the formation of molecular populations in materials is a confusing field of physics with a host of misrepresentation and disagreements. From our vantage point, we are able to discern that the problem arises from a blatant misunderstanding of the space Shelling that presides over the development of matter in our 3-dimensional Universe.

Akin to the particle coupling constant (fine structure constant), the gravitational constant and the electro-static constant, we would expect the field to have established a molecular coupling constant as a pure and simple physical constant setting the scale for the inter-molecular force. The closest thing to such a figure is an *a* constant long established by Danish physicist Diderik Van Der Vaals, which was named after him the *Van Der Vaals constant*. This constant however is not a defining building unit for the crystalline state or more generally the state of Materials as we termed it. Every *structural* material (flat populations of molecules) has been established with its own *a* constant. Furthermore under the umbrella of "Van Der Vaals force" every academic text puts different constructs, some describing it as the total force between two molecules of the material including the repulsive component, others describing it as the total attractive force. Yet others go so far as to apply the defining Van Der Vaals equation to larger macroscopic objects of any kind. What is worse, the officially published list of values of the *a* constant per material, tallies the gas Helium, Krypton and the other noble gases, which are mono-atomic elements, among other multi-molecular gases, demonstrating a complete confusion between state behavior of materials and inter-molecular dynamics in the composition of multi-molecular materials. Although it is fair to say that the original defining equation developed by Van Der Vaals was already a state equation which reads as follows:

$$(p + \frac{n^2 a}{V^2})(V - nb) = nRT ,$$

where a is the inter-molecular force constant. This equation needs an empirical correction, called the Maxwell equal area rule, established by J. C. Maxwell, in order to become thoroughly applicable with sufficient accuracy.

Be all that at it may, it seems reasonable to grant the simplest of all existing multi-molecular materials, the material made of H_2 molecules in any of its states, the etalon of universal inter-molecular force. The constant a for such H_2-H_2 materials, before the current officially established value, was formerly measured to be:

$$a = 0.2444 \text{ liter}^2\text{-atm/mole}^2$$

This value is in close agreement with the Quanto-Geometrically derived value of **0.2441197/σ^3**. The current official value is **0.2476** liter^2-atm/mole^2, not that far from the above. However with the concept now wrongly covering forces between anything from molecular population to macroscopic objects, we will grant better credence and defining stature to the formerly measured value. Furthermore, the term really defining the inter-molecular coupling constant in the above equation is the a term $n^2 a/V^2$, the rest of them being inheritances from state thermodynamics.

In concluding this section we wish to emphasize that the Kinetic Theory based on physical states has to remain separate from the study of inter-atomic attractions or forces, which give rise to the Molecular Shell per our visualization in Quanto-Geometry. The state of materials and their associated kinematics represent a higher level of phenomena incarnated in the inter-molecular dynamics cementing the foundation of the Materials Shell in U_3. Furthermore the Kinetic Theory, despite all its successes, remains in our view incomplete because there exist many more physical states than just the customary three considered (Solid, Liquid and Gases) in accordance with the Quanto-Geometric vertical hierarchy below:

1. $\rightarrow$ *Amorphous Solids*

2. $\rightarrow$ *Crystals*

3. $\rightarrow$ *Powders*

4. $\rightarrow$ *Waxes*

5. $\rightarrow$ *Colloids*

6. $\rightarrow$ *Oils*

7.   → *Aqueous Liquids*

8.   → *Condensed Vapors/Clouds*

9.   → *Volatile Gases*

These definitions are important for such fields as nanotechnology, structural biology, supra-molecular chemistry, polymer science, surface science and perhaps condensed matter physics as a whole. We will further state that these same fields, particularly nanotechnology and structural biology, have much to gain with the experimental application of the physical constants pertaining to the next two Shells, figures thus far unbeknownst to the physical sciences in general. However for these benefits to become tangible, physicists must exercise sufficient intellectual discipline in keeping separate the notion of interaction forces as the building block of a physical Shell from that of state of the population of building blocks making up Shell objects.

5.21. Concordance and Geometry of the Interaction Constants

We shall call our readers's attention to the fact that all 4 of the principal interaction constants belong to the same Quanto-Geometric Hierarchy, a rather remarkable fact attesting to the correctness of the Quanto-Geometric view. What physicists generally hold for the most preeminent constants are those that have to do with the so-called 4 forces: the electromagnetic force, the gravitational force, the strong force and the electroweak force (radioactive). The statute of the strong force, which pertains to both quark interaction within hadron particles and nucleon interaction within atomic nuclides, is ambivalent to say the least, and to that extent may not be classified as foundational as the other three forces at this point. The Quanto-Geometric view offered thru the generating Tri-Valent Hyper Quadratic and Hyper Space Operators offers a harmonized order of these constants because it relates them to their mathematical origin. Therefore the veritable four primary interaction constants are: the fine-structure constant, the gravitational constant, the electrostatic constant and the Van Der Vaals constant as they relate to natural successive Shells of matter in the prime role of forming their building blocks.

This concordance of interaction constants remits us to a different view of matter in U_3 by helping us better understand the Shelling that takes place in the organization of 3-dimensional matter. We have been impervious to that shelling thus far because of our lack of understanding of the nature of space in general despite the advances by Albert Einstein's Relativity. We hope the above study will have made it sufficiently clear that there is a significant distinction, not only between the nature of the scalar element (quantum or mass) in each Shell, given in the specific values of the $M_3(s)$ function itself (under the direction of the Operator), but equally in the multiplicity quality of s_3 space from Shell to Shell, which gives rise to the different forms of intra-shell interactions.

Furthermore it is no less interesting to realize that the equations for all the interaction forces are almost of the same form and all contain a square term in their expression.

Namely:

1. For the Gravitational Force, $F = G\dfrac{M_1 \times M_2}{r^2}$, where G is the gravitational

    constant, M_1 and M_2 are the masses of the objects in interaction and r the distance between them.

2. For the Electrostatic Force, $F = k\dfrac{e_1 \times e_2}{r^2}$, where e is electrostatic constant, e_1

    and e_2 the charges in interaction and r the distance between them.

3. For the Van Der Vaals Force, $F = a\dfrac{n \times n}{V^2}$, where a is the Van Der Vaals con-

    stant, n the total number of molecules and V the volume of the container shared between each molecule.

4. For the so-called Electromagnetic Force, the Force, viewed from a classical vantage point and expressed rather in Power of the so-named Synchrotron Radiation, is captured in this equation:

    $$P = \frac{c}{4\pi\varepsilon_0} \cdot \frac{e \times e}{r^2},$$

    which we have rearranged to make explicit the product and square terms.

The reason why the interaction forces take that very specific form is because they inherit the geometry of the parabolic component of the generating function, which we recall to be:

$$y = \frac{s^2}{4p}$$

If we choose to elect a scalar n to be equal to 2 for the sake of the comparison, the function becomes:

$$y = \frac{1}{p} \cdot \frac{s^2}{n \times n},$$

where n is a scalar of value $n = 2$

$$\text{or } y^{-1} = p \cdot \frac{n \times n}{s^2}$$

Whether we make the comparison directly with y or the inverse function y^{-1} so that the form of the equation becomes exactly identical to that of the Forces is a simple matter of *point of view* from one side or the other. The match stands overly clear in any chosen case.

Accordingly General Relativity makes a fair statement when it typifies the Gravitational force to be a force of *geometric* origin. Just that it is not only the Gravitational force that this qualifier applies to but all the other interaction coupling forces generated by the $M_3(s)$ function and that, in addition, the geometry of this force is preeminently parabolic.

5.22. Quantum Wave Function: Truth or Myth?

In spite of its central role in Quantum Mechanics, no notion has probably generated more controversy in physics than the notion of wave function. Its philosophical implication and consequences insofar as human representation of nature have divided physicists since early-day Quantum Mechanics thru modern-day Quantum Electrodynamics. The representation of the wave function itself, however intuitively factual and consequential it was granted to be, posed enormous interpretative problems. The principle of uncertainty, which springs from an analysis of the modulus squared of the wave function $(\psi)^2$, was the spark that ignited the controversy, with scientists such as Albert Einstein, who himself made contributions to the theory, remaining unswayed by it until their last breath. When Werner Heisenberg proposed that $(\psi)^2$ ought to say something about the localization of the associated particle in space, he was solidifying the way for the mathematical quantum framework that gave spectacular account of the orbital shelling of electrons in the mysterious atom element, as undertaken by the Copenhagen school of quantum physicists and those who subscribed to it. However this aspect of the theory which matured into today's Quantum Field Theory, remains hampered by serious limitations, not the least of which is the inability, on several accounts, to define the wave function of a single photon, as it already was when Einstein, Podolsky and Rosen challenged it with the so-called EPR paradox.

Further, Quantum Field Theory, the latest extension of Quantum Mechanics, has arrived at an impasse when it comes to formulate the granular interaction taking place in the gravitational field on the basis of the quantum wave function, in any attempt to formulate the so avidly sought-after quantum theory of gravitation. In essence, at all orders of the formulation, gravity essentially *gravitates*, and *re-gravitates*, displaying a nightmarish behavior which at the end makes it impossible to segregate gravitational interaction from that of the other forces. Albert Einstein staunchly opposed the notion of the wave function on the basis of being, in his view, a fuzzy abstract mathematical space, and contended that all description of reality should be based on unequivocal and causal description of space and time or "spacetime". He saw no space as it were for the ghost of uncertainty as implied in the principle of uncertainty inherent to conjugate variables of matter. To Einstein's credit, Louis Victor de Broglie, father of the material wave function, did develop in his later years a *real*-valued wave function formalism, in contrast to the common *complex*-valued one. That treatment, however, did not quite reverberate among quantum physicists, both because the complex-based treatment was no doubt more seductive and because it offered a broader analytical framework.

Despite relentless efforts, Albert Einstein could not come up until his last day with the formalism to support his view of the primacy and higher hierarchy of causal relationships over partial, uncertain and probabilistic effects in the description of physical reality.

5.23. Quantum Mechanics's Wave Function Thru Quanto-Geometric Lens

The closest Quanto-Geometric equivalent to the modulus squared of Quantum Mechanics's wave function is the figure:

$$\left[\frac{dQ(s)}{ds} \bullet k_p \right]^2 \text{ or } \left[M_3(s) \right]^2$$

as per Table 5.3. In accordance with the premises that we have set forth in this study, this figure stands for a real quantity. The *s* variable in this figure unequivocally refers to *real* space. The most primal description of the Quanto-Geometric wave function is *s(s*)* as previously stated, *s or s** not being an abstract or complex imaginary but a real physical quantity or an observable despite its nothingness and its infinite "de-ployability". The far-down derivative figure of $\left[M_3(s) \right]^2$ expresses that *de-ploy* ability starting with the index value of s = ½σ or the particle Shell. Effectively a square integer represents the expression of the splitting or de-ploy of a smaller twin quantity: for instance *25* is what *5 x 5* looks like as a square. The squaring operation is a transformation. So therefore the square transform implies the split or conjugation of a quantity that was originally a singlet. Space as a physical variable is the mother of that manner of transform. What's more, the *normal* order inherent to the Quanto-Geometric abstract number stack dictates that the radix quantity (i.e. 5) is smaller in size or dimensioning than its square (25 in this case), but affirmatively larger in scalarity than the square. The Quanto-Geometric function inherits all these properties from the higher number abstraction. In any derivative form where it appears as a composite function, both elements of the composition become individually squared when the whole function is squared since *(ab)²* equals *a² x b²*, which further multiplies the conjugates. This is the prism through which one must understand the modulus squared of wave function in Quantum Mechanics, never to be equated to an abstract mathematical space, once one is able to identify it thru the agency of Quanto-Geometric lens. We further consolidate this interpretation of the material wavefunction in Volume II by deriving the known values of many elementary particles' Compton wavelengths.

5.23.1 Derivation of the Rydberg Energy Constant

The comparable Quanto-Geometric wave function squared is such a real quantity that it takes the majestic incarnation of one of the most *fundamental* physical constants: the *Rydberg energy constant* R_y, with which it presents an otherwise acceptable match. The Rydberg energy constant R_y is the energy of the lowest-energy photon capable of breaking the orbital bond between the electron and the proton that make up the simplest of all atoms, the hydrogen atom. In other words, it is the unit of energy over which the whole atom edifice is generically built across the spectrum of atomic elements, a natural *orbitalization unit*. It is equal to the Rydberg constant R_∞ times *hc* per the following formula:

$$R_y = R_\infty \cdot hc = \frac{m_e e^4}{8\varepsilon_0^2 h^3 c} \cdot hc = 0.136056253 \cdot 10^2 \, eV$$

where *e* is the elementary charge, m_e is the rest mass of the electron, ε_0 is the permittivity of free space, *h* is the Planck constant, and *c* is the speed of light in a vacuum. This is the value published by the National Institute of Science and Technology of the United States in CODATA 2010.

The Quanto-Geometric corresponding value is: **0.1355703/ σ⁶**, portraying an admittedly low figurate match with the currently known value. However if we instead compute the constant from the Fermi coupling constant as an alpha (α) surrogate, and apply the Quanto-Geometric rule of the square of that quantity to produce the orbital constant, the figurate match significantly improves.

From $\dfrac{G_F}{(\hbar c)^3} = 0.116637 \times 10^{-4}$,

we obtain $\left[\dfrac{G_F}{(\hbar c)^3}\right]^2 = 0.136041898 \times 10^{-10}$.

Per our model, the physical constant alpha (α) and the Rydberg energy constant work hand in hand, in that the second is the square of the first: if you don't get alpha desirably accurate, nor will your orbital constant be satisfactorily accurate. Furthermore the above formula for the Rydberg energy constant implicitly shows that alpha (α) does belongs or commonly participate in the computation for the orbital energy constant as part of h^2. We fully uphold and stand by the full figure of **0.1355703/ σ^6** reported in Table 5.3. Modern physics taunts the Rydberg constant R_∞, hence the Rydberg unit of energy R_y, to be the most accurately measured constant ever. The Rydberg constant historically arose as a fitting coefficient in the Rydberg formula for the spectral lines of the hydrogen atom, but Neils Bohr later showed that it could be derived from other more *fundamental* physical constants, thru the relationship reported above, toward supporting his proposed atomic model. From the Quanto-Geometric perspective the Rydberg energy unit R_y qualifies even more as a physical constant than the Rydberg constant R_∞ itself. The assessment of the significance of the square of the M_3 function (for s = $\frac{1}{2}\sigma$) conducted previously implicates the following:

- Orbitals are a result of two elements operating in the form of conjugates of one another within the atomic edifice.

- Every element is appealed to be paired with a conjugate as a norm.

- Each element develops its own orbitalization or shelling (in fact a sub-shelling) in its own right.

- Each orbital may further become or de-ploy into a twin spatial "habitat".

- Every orbital is leveled as a discrete energy pocket scaled on the Rydberg energy unit.

This sets the stage not for a two-tier atom (nucleus-electrons) as traditionally established, but a three-tier atom with two sets of conjugates: electron-proton and proton-neutron. The proton (or protons) is positioned at the mid-tier level and plays an *Ambivalent* role. This architecture not only gathers all the known facts about the atom in terms of electronic distribution and nuclide behavior but has additionally proved able to fully account for all the puzzling *magic numbers* known to the distribution of atomic nuclei of elements, as we have demonstrated elsewhere since 1995 and in Volume II of this study.

5.24. Derivation of the Molar Gas Constant

An ideal gas is defined as a gas whose punctual elements:

- do not interact with each other via any force of any nature.

- are of negligible size with respect to the distance separating them.

- are free to move in any direction and collide only randomly with one another in an elastic fashion.

In the classical approach to this question, the above definition applies to ordinary gases based either on single atoms, specifically the noble gases, or molecules of any composition that are volatile under standard conditions of pressure and temperature (or **STP**).

In order to determine the volume that an ideal gas will occupy in space, three parameters must be taken into account. One is density of the gas or simply the total amount of punctual elements making up the gas. The second is pressure exerted on the sample population and the last one is the temperature of the whole sample. It can be shown that for the same pressure, the same temperature and the same amount of punctual elements, all gases of the ideal type occupy the same spatial volume. Therefore, these three factors, pressure p, temperature T and amount of punctual elements n, may be considered, when taken as a product, as a gas expansion coefficient k. A comparison involving these three factors between two gases will help us gather this fact in a mathematical expression:

$$\frac{p_1 V_1}{n_1 T_1} = \frac{p_2 V_2}{n_2 T_2} = k$$

where p_1 and p_2 are pressures of the two gases, V_1 and V_2 are the volumes of the two gases and n_1 and n_2 the amount of elements in each of the two gases. We may then pursue with a generalization for an ideal gas with the following expression:

$$\frac{p_i V_i}{n_i T_i} = k$$

So that:

$$\frac{V_i}{n_i} = k \cdot \frac{T_i}{p_i} \quad (5\text{-}2)$$

The above equation gives us the inverse density figure of the ideal gas. As a reminder the density is the amount of elements per unit volume. Consequently the inverse of that quantity is the Volume per unit element or the total volume per total amount of elements.

The proportionality constant k is an experimentally measurable quantity and is known as the molar gas constant with the value of 8.3143. If we choose a standard value of pressure (1 at or 101305 Pa) and a standard value of temperature (0°C or 273.15K), the above equation yields:

$$\frac{V_i}{n_A} = 8.1314 \cdot \frac{273.15}{101325}$$

$$\frac{V_i}{n_A} = 0.022413968 \Rightarrow m^3 / mol$$

The change of n_i to n_A is due to the value of the molar gas constant which is based on a specific amount or quantity, the Avogadro number. When we introduce the value of 8.1314 for k then n automatically changes to n_A. The quantity 0.022413968 m³/mol, called the molar volume of an ideal gas, characterizes the space occupancy or localization behavior of an ideal gas. Put differently, every molar amount of an ideal gas will spatial-ize or exert entropy within an occupancy volume of 0.02241 m³. As counter-intuitive as this may appear, this act is played by atoms within the molecular Shell, just as the pre-vious act in this transverse kinematic order was played by particles within the atomic Shell. For that very reason do we see the noble gases, which are mono-atomic elements, equally play this act of spatialization configuration to the same extent as molecular ele-ments. This stands for an epi-phenomenon superimposed on the systematics of inter-atomic bonding that make up molecular elements.

The Quanto-Geometrically derived value for V_i/n_A is **0.0224694201/ σ^6**. The two leading zeros in this case were not introduced by using the device of power of 10 adjust-ment, and must be considered part of the match, making it a 5-digit figurate match. This tight match leads to a remarkable natural validation of the scale of the Pascal and Kelvin units. It also tells us that STP conditions are not an accident on Earth but represent criti-cal points in the development of matter itself.

The high match portrayed by the Quanto-Geometrically derived value of the molar gas constant additionally corroborates the validity of neglecting the possible quantum me-chanical effects at STP. Effectively the conditions that would impose quantum restric-tions on the modality of occupation of the spatial volume do not exist at STP but only at high pressure and high temperature. Nonetheless even in those extreme conditions, the basic gas law relationship (5-2) would not totally collapse, it would only change by an understandable proportionality constant, as in effect it does, since the relationship simply goes:

$$\text{from: } p_i V_i = nkT$$

$$\text{to: } pV = \frac{2}{5} nkT.$$

The latter is the relationship that describes the spatial occupancy of the Fermi Gas, a so-called gas of electrons. In those circumstances the departure of the gas behavior away from the classical gas constant anchor is completely understandable, since it constitutes no more than a regression to the atom and particle Shells catalyzed by the shorter scales and high energy levels common to these lower habitats.

5.25. The Cosmological Constant and the Anthropic Principle

A discussion of the Cosmological constant seems to be in order in this text given the endless talk over the last decade regarding that figure in the field. The terms of the discussion is however completely ill-framed from our point of view, since the predicates for this figure have been all but unknown to this point.

What the field currently refers to as the cosmological constant is what is called the "vacuum energy density", which it assumes to be the same wherever you look at the vacuum in the universe, whether it addresses the vacuum between galaxies or the vacuum between atomic elements for instance. Further there is a whole series of considerations and hypotheses that arise from the value of that constant in terms of the configuration of U_3, particularly the question of whether or not the universe is expanding, contracting or static. This constant seems to have been ill-fated since its inception by Albert Einstein in the first half of the 20th century, because he devised it as a figure to justify the stationary universe known at the time when shortly thereafter observational evidence pointed in the direction of an expanding or inflationary universe instead, the "greatest blunder" of his life in his own words.

Discussing a cosmological constant as the generic "energy of pure vacuum" without understanding the Shelling hierarchy based on the states of entropy inherent to each Shell, which give birth to the host of physical constants, is a complete act of blindness. The only kinematic characterization in U_3 that applies as a generalization to all vacua, throughout the integral suite of Shell deployments, is the propagation speed of space, which the field calls the speed of light. Beyond that figure, every Shell of space vacuum must be treated individually. There is no "cosmological constant" that universally applies to the larger-scale galactic groups, galaxies, solar systems and so on and so forth. There is only a specific constant for each of these Shells per chart in Table 4.3.

The proponents of the notion of cosmological constant realize that there must be a connection between the physics of an expanding universe thru cosmic vacuum and the physical constants that relate to the vacuum at lower scale. In their view, the vacuum at larger scale is the same as the vacuum at lower scale. So therefore an expanding vacuum at higher scale must imply an expanding vacuum at lower scale as well. This poses the nagging question of the consequences of an expanding lower-scale vacuum over the physical constants, since the vacuum state governs those lower scale physical constants. Another implication is that the physical constants may not only remain constant over time but most significantly may not retain their exact same value everywhere in the universe.

There are a number of consequences to these views. One leads us to the conclusion that as a result of an expanding lower-scale vacuum all things humans are familiar with in their habitat must be expanding in size, including themselves, either in a uniformly proportional fashion or in a non-linear fashion depending on which physical constants get affected the most. This could mean that if a human were to live thousands or tens of thousands of years (or simply applying the concept across a large enough generational span) at end of lifecycle (or the generational span) a human might not be able to get into or exit his or her home because of disruptive size expansion of matter between beings and habitat. The proponents of the idea of universal inflationary expansion of the vacuum will discard these contentious consequences in our sense by proposing a so-called Anthropic Principle (*anthro* means human), among other reasons inherent to their theory. According to the Anthropic Principle, constants of nature are not universal in the cosmic universe and take different values in different regions of the universe. Only certain regions of the universe are fit for human life in suitable material habitats, ours being one of them! This certainly has the likes of these short-handed, arbitrary and self-serving methods and misnomers that so much upset Paul Dirac. Without any further comment, the Anthropic Principle qualifies as no less than an unsound principle and should summarily cease and desist!

So should all the ideas that drive the Anthropic Principle. The arguments that we have put forth in order to refute the possibility of alteration of the fundamental physical constants in the next Chapter of this study also hold against variation of these constants as a function of locality in U_3, and consequently they equally hold against the Anthropic Principle. After taking stock of that development the reader may conveniently revisit this section.

A cosmological constant at larger-scale structure means **not** an expansion of space such that each and every galaxy will ultimately find itself absolutely lonely in a rarefied and deserted Universe. The cosmological constant should be understood instead within the framework of the $M_3(s)$ function and interpreted within the transverse hierarchy of localization. The Hubble law (based on the Doppler effect) does not mean that because galaxies are receding from one another, the vacuum they populate is in inflationary expansion, just as two electrons in an orbital constantly receding from one another do not describe an inflationary orbital vacuum. The Hubble law must be interpreted along the line of localization instead of inflationary expansion, while the cosmological constant will have better success within the interpretative framework of punctual population statistics, otherwise known as Fermi-Dirac statistics, Bose-Einstein statistics, and even the classical Kinetic Theory, all of which have proven very successful and adequate theoretical frameworks.

The current computed value of the cosmological constant is given by a variation of the formula proposed long ago by A. Einstein:

$$\Omega = \frac{8\pi G}{3H^2 c^2} \cdot \rho$$

where G is the Newton Gravitation Constant, H is the Hubble constant, c is the speed of light and ρ is the vacuum energy density. After careful and repeated examination of the numbers deriving from this formula in several different dimensions, we could not find any match with any of the constants marked *unknown* in the astrophysical Shells of our model chart. Therefore we summarily dismiss these Ω numbers as fundamental as we further discuss them in greater detail below. We believe that all the historic cosmological data accumulated to date must be put back on the table and carefully re-examined in the light of population statistics, an exercise that might yield measurable quantities with appreciable matches with the Quanto-Geometrically derived astrophysical constants for their corroboration.

5.26. Quanto-Geometric Cosmogony

The cosmogony that directly emerges from the Quanto-Geometric framework, in particular the Grand Eigenfunction at the heart of the Theory, is a Universe that appears as an *Omniverse*, with 9 basis adiabatic layers or strata in the vertical order, which we have designated World Planes, and in the horizontal order a transverse span of Shells, in principle of boundless extension despite their "nanoscale" or 9-dimensional bundling. We acknowledge the hypothesis that in the vertical order the Omniverse could be bundled as well with a global boundless span.

The World Planes are meta-physical to one another while characterized by their specific degrees of freedom dictated by the primeval $\{s_n\}$ scaling sequence. This sequence defines the numeric value or Modulus of the dimensional set that makes up the background of each of these World Planes. Despite their tight mutual estrangement, their adiabatic partition is not absolute, since the mere virtue that a World Plane agent or observer may envision or formulate vertical trans-ontology is in itself a testimony to inter-penetrability of some sorts, at least at the consciousness level, which is a fact of fully physical nature. One other significant reason is that the σ standard deviation metric is common to all World Planes of the Omniverse.

We intuitively recognize that the Physical Universe that we inhabit as agents and observers is endowed with 3 degrees of freedom. The question of the derivation of dimensional set $D = 3$ has been posed by physicist Martin Rees as a non trivial question deserving serious attention. In other words, the question as to why our universe is 3-dimensional is a rightful question to pose to human intuition or common sense. We have already explained how the background in this Universal Arena of ours is constructed, that $D = 3$ is actually $D_3 = [2+1]$, we now have to add that the reason why there is a 3-D World Plane is because $n = 3$ belongs in the range of the $\{s_n\}$ sequence which defines the several dimensional sets of the Omniverse. We, as intelligent beings able to pose questions such as these, just happen to occupy the 3-D World Plane. The idea that the other World Planes of different dimensional moduli do not host intelligent agents is unfounded from the Quanto-Geometric viewpoint, because the transverse span of Shells that see matter develop from the infinitesimal quantum to the karyotic forms of the living to the cosmological heavens is common to all World Planes. The adiabatic division that guarantees the independence of each World Plane and sustains the laws of symmetry on which each is built understandably prevents at the same time all forms of direct inter-element interaction.

5.27. Vacuum Energy Density and Standard Deviation of the Omniverse

Gathered Supernova data in the field of astrophysics suggest that the vacuum energy density or "zero-point mass density" of the 3-D universe is greater than zero. From computation focusing on electromagnetic force, one obtains the not so surprising result that empty space in the 3-D universal arena "weighs" 10^{93} *grams per cc*. This result would be quite significant if true. However, the actual average mass density of the 3-D universe is computed to be 10^{-28} *grams per cc*. Cosmic background radiation (CMB) data combined with the measured Hubble constant do confirm the supernova data that suggests a value of 10^{-28} *grams per cc* for the average mass density of the universe. Astrophysicists believe that mass density is equivalent to the vacuum energy density per current understanding in the field.

This discrepancy between expected value and measured value, in the order of 10^{-120}, has created a deep decade-long crisis in the field of cosmology. Astrophysicists find themselves faced to the conundrum of explaining the astonishing fact that the positive and negative contributions to the "cosmological constant" cancel to *120-digit* accuracy, and yet fail to cancel beginning at the *121st digit*. This paradox only comes to corroborate the problem with the visualized nature of the "cosmological constant" as argued previously.

We shall duly notice the coincidence in pure magnitude between the average mass density computed from cosmological data and the primal standard deviation parameter cubed for a quantum-space distribution applying to the 3-dimensional universe, σ^3, that is. If the adjusted nanometer constant indeed represents the physical materialization of σ for the Omniverse, which would equally apply to the 3-D World Plane, then:

$$\sigma^3 = \left(1.00001203 \cdot 10^{-9}\right)^3$$

$$\sigma^3 = 1.00003609 \cdot 10^{-27}$$

The magnitude of σ^3 is in very close agreement with the 10^{-29} quantity for the vacuum energy density in the 3-D World Plane, as subsequently reported, showing a difference in only 2 orders of magnitude. Furthermore, the intrinsic dimension of the number (m_E^3 , wavelength of harmonic fluctuation in cubic meter) is in close agreement as well with the λ_C^3 intrinsic unit-dimension (Compton wavelength in cubic meter) of the vacuum energy density figure.

What's more, if the transversal hierarchy of Shells in the 3-D World Plane surrogates in any way the 9-stratum vertical hierarchy of the Omniverse, as suggested earlier in this study, it should not be any surprise that there is 3 different possible phases for the large scale geometry of the 3-D World Plane as currently hypothesized in astrophysics. This triplet latency is in complete accordance with the Quanto-Geometric Tri-Valent characterization. Effectively, if we use Ω_Λ to denote the critical limit density and Ω_M to denote the ratio of ordinary matter density to critical density, then:

1. The 3-D World Plane is open if $\Omega_M + \Omega_\Lambda$ is less than one (space-dominant).

2. The 3-D World Plane is closed if $\Omega_M + \Omega_\Lambda$ is greater than one (scalar/mass-dominant).

3. The 3-D World Plane is flat if $\Omega_M + \Omega_\Lambda$ is exactly one (scalar-mass and space covariantly equivalent).

The ratio of ρ_{vacuum} to $\rho_{critical}$ is in offered calculations what is denoted Ω_Λ, expressing the vacuum energy density on the same scale used by the density parameter Ω, leading to the relation:

$$\Omega_\Lambda = \frac{\rho_{vacuum}}{\rho_{critical}}$$

The supernova results have suggested that the vacuum energy density is close to the theorized limit:

$$\rho_{vacuum} = 0.75 \cdot \rho_{critical} = 6 \cdot 10^{-29} \, gm / cc$$

Thus the data suggests that $\Omega_\Lambda = 0.75$, and that therefore the 3-D universe is in constant expansion. These views are encapsulated in NASA's depiction of the evolution of the 3-D arena reported in Fig.6.2, which interestingly reflects a quanto-geometric distribution. However, from our standpoint, the sketch should have been drawn as a probabilistic distribution instead of anchored on the usual timeline. That graphical depiction alone should tell astrophysicists that beyond the representation of a putative evolution in time for the 3-D World Plane lies indeed a larger normal probabilistic distribution, which is essentially timeless.

Note further that the current view is that if Ω_Λ is greater than zero, then the Universe will expand forever unless the matter density Ω_M is much larger than current observations have suggested. For the Ω_Λ greater than zero scenario, even a closed Universe would expand forever.

It is important to put in highlight that the above reported numeric details reflect the cumulative inquisitive thinking in cosmology today about the dynamics of the *observable* universe and only enjoy limited support from the sketched Quanto-Geometric perspective. The more fundamental view remains the true cosmological constant figures deriving from the framework exposed in the fundamental physical constant chart and their dynamical implications in terms of the behavior of the entire 3-D World Plane which is not altogether observable.

5.28. Scalar Fields in the Standard Model

The treatment of sup-particle dynamics we defer to the next cast of this study, as previously advised. However, due to the nomenclature being used lately in reference to the latest findings about bosonic scalar fields, such as related to the Higgs boson, we feel compelled to bring a few conceptual adjustments to the discussion. In addition, the notion of "pure scalar fields" in terms of a mechanism to generate mass for elementary particles seems to be in contradiction with the Quanto-Geometric axiom that "pure scalar" cannot exist, per our contention at the beginning of this study, on the basis that a minimum level of vacuum component is a requirement for the physical inception of a scalar. This all begs a clarification.

Let us note, at the outset, that the treatment of "pure physical scalar fields" is entirely based on the mathematical model for scalar fields. A mathematical scalar field is a region U defined by a complex function, distribution or matrix, admitting points with complex coordinates (real and imaginary) where the real part is an invariant independent from the field region and its coordinates, and the imaginary part is fully dependent on the spatial nature of the region or field. The field and the populating scalar form a complex conjugate while the topological point is equally a complex quantity in itself. If the field is a physical tensor or gauge field, it may thus provide scalarity to the point(s) of population by means of interaction with these points thru their imaginary quantity. One can see that this analysis is based on the type **5 Primitive** of the Higher Numeric Abstraction, the realm of complex numbers. Therefore the "pure physical scalar field" that the Standard Model refers to is merely an idealization of a pure scalar, which does not even qualify as a *monolith*. It is at best a form of scalar, a lower level entity in the scalar hierarchy.

What's more, the mere fact that a scalar packet can be broken by a highly energetic projectile (as it happens in high energy colliders) means that:

1. The packet was a dimensional entity, with a vacuum component that is, which provided the means for the segregation (a pure monolith is conceivably unbreakable).

2. The dichotomized free particle, i.e. a Higgs boson, is still a dimensional element because it can be subsequently observed or accounted for, for it would otherwise vanish as a dimension-less element.

From that standpoint, the idea of mass as a "pure scalar" is untenable and is only sensible as an idealized pure scalar. In general, scalar field theory does not precisely explain how mass is "created", shall we say, but how mass is induced or compounded thru the Lagrangian transform of symmetry breaking that relates the imaginary component of the elemental packet to the field. So it is since an invariant (tenet with mass dimension) must pre-exist in the quality of the elemental packet for the action or the Lagrangian transformation to take place as a paramount condition for the operation. Philosophically speaking, there is nothing in abstraction or in existence that can be innately motionless other than a very scalar.

It is worth noting that we have also made serious advances in the quest for allocation of mass to elementary particles with the direct derivation of the value of interaction coupling constant α, the Fermi coupling constant and the Rydberg constant. It is important to recognize once again that in neutron decay, where the Rydberg constant intervenes, the first particle emanating from degeneration is the heavy W Boson. The energy of the Rydberg constant is also the energy or mass of that boson, the value of which was Quanto-Geometrically derived.

In the aggregate, we remain highly skeptical of "pure scalar fields" or "dark matter" in the cosmological heavens in connection with the cosmological constant, as hypothesized in the field, having seen that the phenomenon of mass allocation is essentially orchestrated within the S_0 protoShell. From a cultural point of view, there is more to gain in the manipulation of entropy in the higher Shells than the manipulation of elemental scalars in the S_0 protoShell. Simply because the deconstruction of the foundations leads to infernal chaos whereas the ubiquitous propagation of the activity under the Sun and beyond is glory and majesty.

5.29. Graphical View of the U_3 Creation Function

Toward performing the graphing of the Creation function, we already know of nine critical points of this graph, which are the points that produce the transverse hierarchy of interaction coupling constants.

A.- To further facilitate the graphing, we will compute the $Q(s)$ coordinate of the function at $s = \frac{1}{4}\,\sigma$, as completed below.

$$M_x(s_x) = 0.66114378278 \cdot \frac{-s}{\sigma^4\sqrt{2\pi}} \cdot e^{\frac{-s^2}{2\sigma^2}}$$

$$= 0.66114378278 \cdot \frac{-(\sigma/4) \cdot e^{\frac{-(\sigma)^2}{16 \cdot 2\sigma^2}}}{\sigma^4\sqrt{2\pi}}$$

$$= 0.66114378278 \cdot \frac{-1 \cdot e^{-1/32}}{4\sigma^3\sqrt{2\pi}}$$

$$= 0.66114378278 \cdot \frac{-0.096667029}{\sigma^3}$$

$$= \frac{0.063910805}{\sigma^3}$$

We are now going to inquire about the minima and maxima of this function with the zeros of its derivative. The function is:

$$M = \frac{dQ(s)}{ds} \bullet y_p = \frac{-s}{\sigma^3\sqrt{2\pi}} \cdot e^{\frac{-s^2}{2\sigma^2}} \bullet \frac{\sqrt{7}}{4\sigma}s^2$$

$$= e^{\frac{-s^2}{2\sigma^2}} \bullet \frac{\sqrt{7}}{4\sigma^4 \sqrt{2\pi}} s^3$$

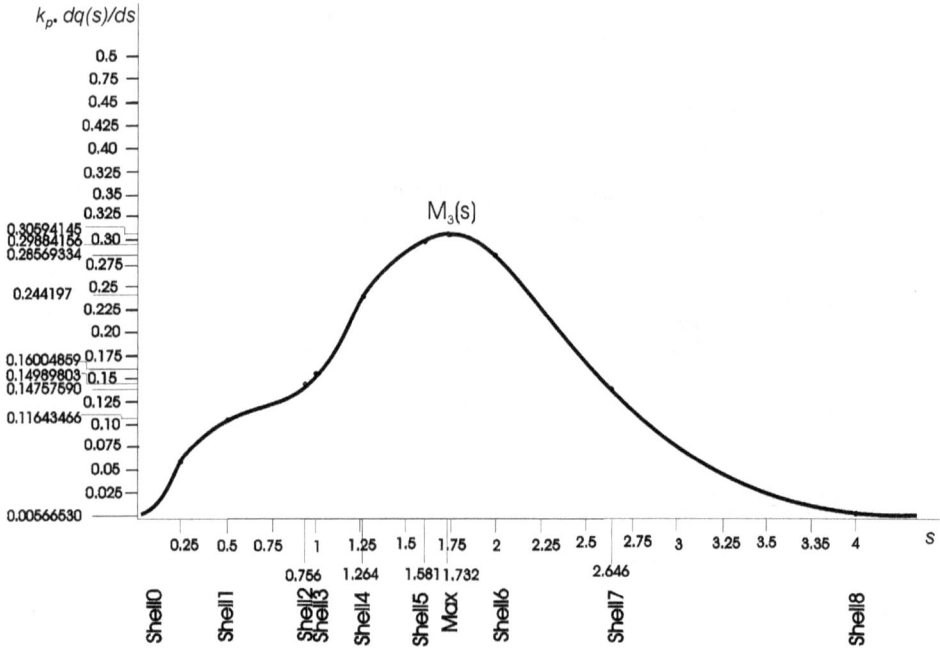

k_p. $dq(s)/ds$

0.5
0.75
0.45
0.425
0.40
0.325
0.35
0.325
0.30524145
0.29884156 0.30
0.28569334 0.275
0.244197   0.25
0.225
0.20
0.16004859 0.175
0.14989803 0.15
0.14757590
0.125
0.11643466 0.10
0.075
0.05
0.025
0.00566530

0.25 0.5 0.75   1   1.25 1.5 1.75   2   2.25 2.5   2.75   3   3.25 3.5 3.35   4   S

0.756   1.264   1.581 1.732   2.646

Shell0   Shell1   Shell2 Shell3 Shell4 Shell5 Max Shell6   Shell7   Shell8

$M_3(s)$

Fig. 5.11 Graph of the $M_3(s)$ creation function

Let's segregate the coefficient into:

$$k = \frac{\sqrt{7}}{4\sigma^4 \sqrt{2\pi}},$$

$$\text{so that } M = ks^3 e^{\frac{-s^2}{2\sigma^2}}$$

The derivative of M is:

$$\frac{dM(s)}{ds} = k\left[3s^2 \cdot e^{\frac{-s^2}{2\sigma^2}} + \frac{-2s}{2\sigma^2} \cdot e^{\frac{-s^2}{2\sigma^2}} \cdot s^3 \right]$$

$$\frac{dM(s)}{ds} = k\left[3s^2 \cdot e^{\frac{-s^2}{2\sigma^2}} - \frac{s^4}{\sigma^2} \cdot e^{\frac{-s^2}{2\sigma^2}}\right]$$

There are two possible conditions to be met for $\dfrac{dM(s)}{ds} = 0$.

The first is $k = 0$, which is not a possible or meaningful condition.

The second is for $\left[3s^2 \cdot e^{\frac{-s^2}{2\sigma^2}} - \dfrac{s^4}{\sigma^2} \cdot e^{\frac{-s^2}{2\sigma^2}}\right] = 0$

$$\left[3s^2 \cdot e^{\frac{-s^2}{2\sigma^2}} - \frac{s^4}{\sigma^2} \cdot e^{\frac{-s^2}{2\sigma^2}}\right] = 0$$

$$e^{\frac{-s^2}{2\sigma^2}}(3s^2 - \frac{s^4}{\sigma^2}) = 0$$

This implies: $e^{\frac{-s^2}{2\sigma^2}} = 0 \rightarrow s = \infty$,

or $3s^2 - \dfrac{s^4}{\sigma^2} = 0$.

Which in turn implies: $s^2(3 - \dfrac{s^2}{\sigma^2}) = 0 \rightarrow$ **s = 0**, or $\dfrac{s^2}{\sigma^2} = 3$

$$s = \pm\sigma\sqrt{3}$$

Therefore we have 3 extrema: **s = ∞, s = 0** and $s = \pm\sigma\sqrt{3}$.

B.- Let's compute the q coordinate for that last abscissa from $M(s)$:

$$e^{\frac{-s^2}{2\sigma^2}} \bullet \frac{\sqrt{7}}{4\sigma^4\sqrt{2\pi}} s^3$$

By replacing s with $\sigma\sqrt{3}$,

$$M = \frac{0.3059421073}{\sigma}$$

So therefore, that last point is a maximum, and M experiences minima at $s = 0$ and $s = \infty$. However we must remember that the Function is undefined at $s = 0$.

C.- We will also study in this section the indefinite integral of the $M(s)$ function. To do so we will employ the so-called method of integration by parts as shown below. Let:

$$T(s) = \int M(s) \cdot ds = \int -ks^3 e^{\frac{-s^2}{2\sigma^2}} ds$$

Let $u = -ks^2$ and $du = -2ks \cdot ds$

Let $dv = s \cdot e^{\frac{-s^2}{2\sigma^2}} \cdot ds$, from which $v = \int dv = -\sigma^2 \cdot e^{\frac{-s^2}{2\sigma^2}} + C$

$$T(s) = uv - \int v \cdot du = (-ks^2)(-\sigma^2 e^{\frac{-s^2}{2\sigma^2}}) - \int (-\sigma^2) \cdot e^{\frac{-s^2}{2\sigma^2}} (-2ks) \cdot ds$$

$$= k\sigma^2 s^2 e^{\frac{-s^2}{2\sigma^2}} + 2k\sigma^4 e^{\frac{-s^2}{2\sigma^2}}$$

$$= k\sigma^2 e^{\frac{-s^2}{2\sigma^2}} (s^2 + 2\sigma^2)$$

Replacing k by its value:

$$T(s) = \frac{\sqrt{7}}{4\sigma^4\sqrt{2\pi}}\sigma^2 e^{\frac{-s^2}{2\sigma^2}}(s^2 + 2\sigma^2) + C$$

$$T(s) = \frac{\sqrt{7}}{4\sigma^2\sqrt{2\pi}}e^{\frac{-s^2}{2\sigma^2}}(s^2 + 2\sigma^2) + C$$

This is an important quantification integral to which we will devote more attention in the next cast of this study.

5.30. Role of Mathematics in Physics and State of Physics Today

The purist and trained physicists may not have noticed in the above developments much of the algebras, the Lagrangians and other complex forms of mathematical structures that they might expect to mandatorily comport the agency of such foundational physical dynamics. This is a demonstration that much of it is useless for the inquest of fundamental constants in foundational physics. I am certainly not about to apologize for Nature preferring to be simple, clear and straightforward. Further, we have expressly chosen to remain clear of all complicated formalism, preferring to adopt as simple mathematical structures and language as possible in order to keep this text amenable to the most universal audience.

In addition, we do wish to bear witness that good physics, otherwise described as a rigorous and coherent description of nature, may be done with no recourse to and seeking no legitimacy in overly complex mathematics, a long-running problem in modern physics acknowledged to have been acutely sterile in results since the 1970's. The abundance, elegance and complexity of mathematical developments, all of which goes much to the enjoyment of the sanctimonious cast, can never substitute the formulation of clear and proper abstraction correctly positioned to ultimately expel the final solution to the problem at hand. Mathematical developments with no direct and tangible results in the resolution of a physical problem remain but mathematics.

The preeminence of abstraction that we spoke of in concluding Chapter 3 should not be construed in terms of a free license to ill-fated physics Analytic models should not seek comfort in intense and sophisticated mathematics in the lack of resolution of physical conundrums they are meant to elucidate. The higher abstraction is and must be the realm that gives rise to the elements of solution to the problem, it can never lay in segregation with the problem in question, and further ought to provide a clear path to its resolution, lest the abstraction becomes useless and almost meaningless aether. What we have argued is that the number stack represents the highest level of abstraction that models physical reality as a whole, but it only makes sense to us strictly in the context of physical reality and the unraveling of its mysteries, never as an end to itself. In that sense mathematics are abstract solutions awaiting to be applied, the exercise being frankly totally dispensable and in the long run valueless if it can never meet physical applications. Of even more contestable value is the physics that does not produce physical solutions but only new provinces within mathematics, which is not at all its mission.

Background Correspondence

	Quanto-Geometry	Other Systems
Philosophy	Quantum-Space systematics	Dialectic Idealism
Mathematics	Theory of Hierarchisation and Typification of the Number Stack	Traditional unspecific Number Theory
Physics	Coupling effects of space over mass	Quantum Scalar Fields in Quantum Field Theory and Standard Model
	Coupling effects of mass over space	Conversion of mass into energy in Special Relativity
	Complete dismissal of time variable	All abstracts based on time except in Quantum Mechanics
	Horizontal Normalization resulting in 9 Phenotypical Primitives or Typified Certainties	Vertical Normalization resulting in a probabilistic spread or certainty-uncertainty conjugate
	Real-image Spinor space for elemental objects	Abstract Spinor space in Quantum Mechanics and Standard Model
	Derivation of physical constant quantities	None - Properties of matter in abstract only
	Representation of Grand Universe staged on 9 real World planes	Representation of 11 Universes, with only D4 real, in String Theory

Table 5.5 Conceptual correspondence between Quanto-Geometry and other Systems

Finally, a brief press review of the current state of the discipline will be quick to reveal a pervasive discomfort at all levels in regards to the long-running squint level of productivity of the field. Lack of new and plural ideas is a matter of concern, overt in some circles and reportedly camouflaged in others, while most of the few ideas enjoying prevalence nowadays stand in many ways for recycled notions that the great generation of physicists of mainly first half 20th century credited with no or little merits. Most importantly the new and bold ideas remain no less than unconfirmed hypotheses for the most part. This is not an attempt to discredit any one physicist of the day or any work. Indeed the *trouble with physics* today, as some have put it, goes in our estimation much beyond physicists, despite the above critical remarks to their address, and stems from the prevailing values in the culture, which negatively reflects on institutions and ultimately on physicists and their practices as well. The culture of foremost mercantilism and blind gain, in almost complete detriment of expression of the humanities, is a societal problem that, at its recently attained peak, wrecked havoc of transatlantic and southern economies on the planet, and with them many human lives, notwithstanding the many disingenuous interpretations to the contrary by these actors. In this landscape, the career of a physicist is ultimately faced with increasing institutional pressures that end up suffocating the vibrancy and productivity of the field. This problem is not going to be solved in our physics conference rooms and fora. It is a severe societal problem that requires for its resolution a push for a larger degree of freedom of the humanities vis-à-vis mercantilism, both of which dually and naturally configure the Quanto-Geometric engine that orchestrates societal dynamics.

Lastly, we offer in Table 5.5 a brief parallel between high-level Quanto-Geometric constructs and corresponding constructs in other physics theories and philosophy, which many may find useful to a final reconciliation of ideas.

5.31. Conclusion

In concluding this Chapter, we shall point out that all of the unknown constants listed in Table 5.3, although undiscovered thus far, are completely existent physical quantities in their own rights, to the same extent as the other well-known constants we have just studied. While ideas about the cosmological constant continue to be actively debated and remain in flux, a *life* constant is nowhere under the prospective radar of academia, the assumption in most quarters being that life with its bewildering complexities is probably not submissive to mathematical determinants. The Quanto-Geometric prediction of a physical life or *karyotic* constant is deemed to profoundly alter, if not revolutionize, our idea of the *living* by placing it in the context of the transverse hierarchy of Quanto-Geometric *de-ployments* in the 3-dimensional universe.

Yet the reader may wonder at this end point where is the debate and argumentation in favor of the meta-physical realm promised at the introduction? Well, there is no debate, we have let the exposition speak for itself: the physicality of the meta-physical realm is an inherent property of the entire Quanto-Geometric functional framework; its unraveling of the pillars on which the physical U_3 universe is erected stands for an undismissable validation of the constructs it purports in relation to all predicted U_i meta-physical universes of the Omniverse. The higher transverse Universal hierarchies on the vertical line we have called World Planes are as real as the 3-dimensional transverse hierarchy in the 3-D World Plane. Just as we have seen that there exists a complete program of organizational instantiation of the fundamental constants upon which U_3's Shells are built, there exists a similar program for the foundational organization of every other World Plane of the Omniverse. The repeated attempts that we have seen in theoretical physics in describing multi-dimensional space (or spacetime as ordinarily denoted) may be taken as efforts in that direction but remain unfounded and frankly quite mislaid due to their total lack of connection with the fundamental constants of Nature. It is my personal view that Religion of any revealed strain, well understood, is Cosmic Science. End of discussion.

Finally, it is well worth noting the following remarks by E. Schrödinger for an introduction to his 1950 publication *Structure of Spacetime*:

In Einstein's theory of gravitation, matter and its dynamical interactions are based on the notion of an intrinsic geometric structure of the spacetime continuum. The ideal aspiration, the ultimate goal, of the theory is not more and not less than this: A four-dimensional continuum endowed with a certain intrinsic geometric structure, a structure that is subject to certain inherent purely geometric laws, is to be an adequate model or picture of the"real world around in space and time" with all that it contains and including its total behavior, the display of all events going on in it.

It seems to me that the Quanto-Geometric Theory, in the entire developments that we have hereby undertaken, furnishes for its own a vast response to this goal of old by successive generations of physicists of an encompassing generalized description of Nature from the inner qualities of ubiquitous and universal space. Except that it is cast far and away from *time*, of course.

Chapter 6

*Time is what Humans measure
on clocks!*
(SIC)

THIS IMPOSTOR OF TIME

If time is what humans measure on clocks, we then have an absolute, since it has been raised to the status of a physical dimension, that is dependent not only on humans but on their clocks as well. This is the most absurd definition ever proposed, it makes a complete disservice to the reputation of human intelligence in this corner of the woods we call the Universe. The notion of time in our minds is a very resilient and tenacious abstraction that has accomplished the feat of resisting the assaults of Quantum Mechanics, which has completely demoted and unseated it insofar as the visualization of one of the physical realms. No one seem to wonder why a whole span of the universe would be so disparate and disjunctive that at one scale it would be entirely bereft of a "universal" variable called time, incidentally its very foundation inasmuch as atoms and sub-atomic particles constitute for the most part its bedrock structures, and at a higher encompassing scale it is completely anchored on this variable. A visualization which further implicates that this variable has started to have a life since the very foundation when the particles themselves were thrown into existence. This stands for the all brave contention of the Standard Model's Big Bang Theory about the formation of the three-dimensional universe! If time has started to elapse since the initial Big Bang event, why did it suddenly disappear at the atomic and molecular phase of development and suddenly re-appear going forward into the development of the following structures? Is our Universe anchored on a shaky and broken timeline? This picture is very seriously flawed, to say the least. We shall actively continue henceforth our crusade against this false and baffling notion of time. We will also revisit in this Chapter other crucial concepts such as evolution of matter and life as well as the expectations the field has assiduously harbored for a more consistent visualization of matter at all scales, this in the light of the success of the Quanto-Geometric constructs in unraveling the mathematics behind the physical constants.

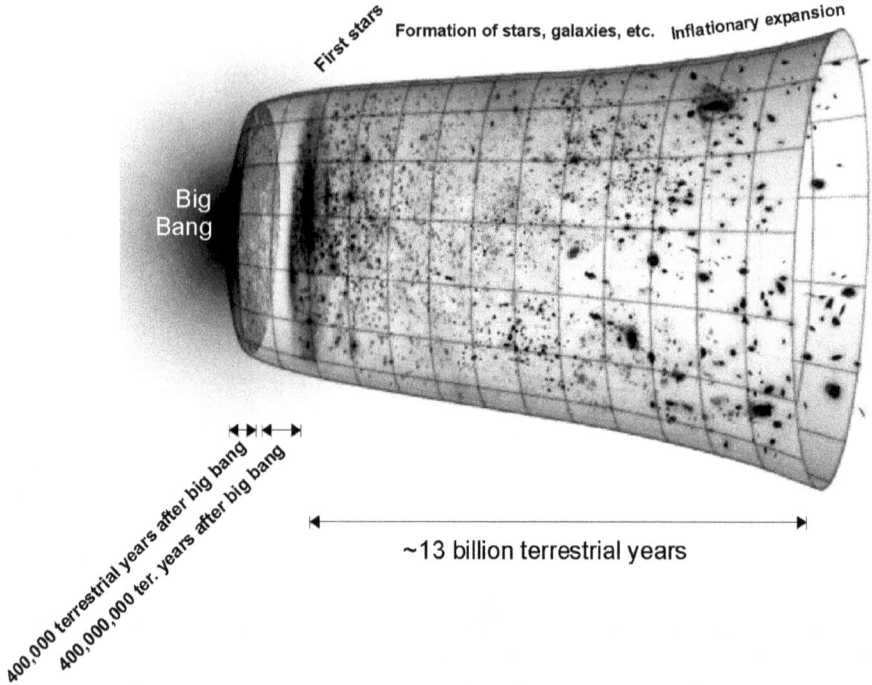

Fig. 6.1 Standard Model's Big Bang Universe

6.32. Apologies But No Big Bang

The observation that the Universe is in expansion in all directions is in direct contradiction with the idea that it has started from a singularity or one single point explosively (Fig. 6.1). If the universe has started from one single point, at a time t_0, then all subsequent expansion must have developed along a timeline starting at t_0 from one point s_0 in space. That point s_0 must have existed even though it could no longer be observable. We would have a sense of where it lies because the "spacetime" arrow would develop from it. In such a scenario all matter would be expanding along virtual cones centered about virtual s_0 as the center of a virtual sphere of expansion. This would give us spacetime with a number of timelines but all equated to one another and all collapsible into one. Therefore the scale of time would always be the same to all observers within the universe.

The Theory of Relativity has already broken apart the visualization of one universal time scale to all observers. But since the construct of time in our minds will not die, among some of the ideas that followed suit since then is the notion that, at early times of the universe, travel light has created space, another absurd construct. The late John A. Wheeler, who had been instrumental in the development of these ideas, hypothesizes:

Maybe the universe itself sprang into existence out of nothingness, a gigantic vacuum fluctuation which we know today as the big bang.

Probably a best-effort yet oblivious visualization by an otherwise brilliant and creative mind to frame the bigger picture. Unfortunately the bigger picture admits no frame and no framing as a condition for successful *in-sight*.

All of these concepts are in truth *monocentric* constructions corresponding to a quantum-dominant form of culture that has been prevalent in the West. We want to find at all costs a *beginning* to the universe, a *singularity* from which it emerged because this is the monocentric *way* we think (Fig. 6.2). The astrophysicist who wants to find a single-point origin to the universe is thinking no different than the religious spirit who wants to find a single God to be the Creator of the universe. Furthermore the monocentrism goes so far as to stipulate that men must be at the image of this God. The scientist wants the universe to have surged from a single point as much as the religious wants it to have surged from a point-like God as the single primal cause of all things in existence. If God had intelligence, a know-how whereby he could put in existence a world of differentiated things, he must have been himself a very differentiated, complex and organized structure (because in our experience intelligence is complex), something like a self-replicating structure from which the act of creation proceeded. If so, then we are back to the beginning, since the question becomes again: where does He come from, what singularity has given birth to Him? The proponents of the Big Bang singularity are quick to close this endless loop by cleverly stating that t_0 is the farthest instance of time that will ever make sense. Yet I must insistently ask Mr Big Bang, what of You had exploded at time t_0, where does it come from and where is s_0 today?

If we were to look at the Universe from a space point of view instead, we would understand that the hypothetical scalar that exploded at time t_0 could not be an absolutely pure scalar but a composite scalar-space entity. It could only have exploded by virtue of sudden inner space expansion. From the fullness of **type 9 Primitive**, we can appreciate that space or nothingness cannot be a progeny of anything *else*. It is not laid in points, but in infinites. Therefore there is no single point from which it comes. It does not and cannot come from some-*where*, because it is the *where* of itself and all *where* is of its own. Consequently the only sensible conclusion is that it has *always* been *there* and has always been in expansion if ever. Despite its possible constant expansion, there is nothing beyond it, because every "where" we can think of it is "there", in the infinites. The movements and transformations that we may witness in the observable universe

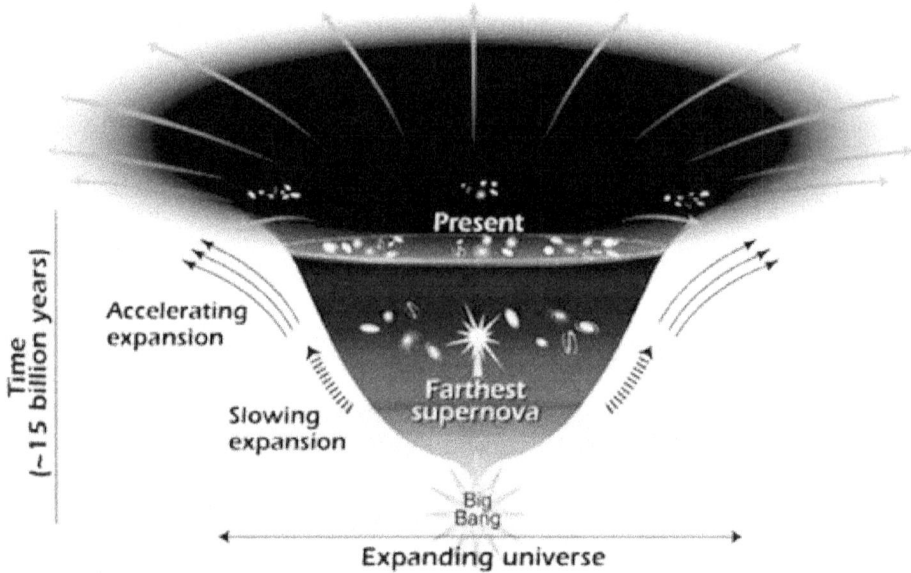

Fig. 6.2 NASA's model of an inflationary universe

are **locally instigated by partitions of space** that materialize the triplet cycle of life-death-rebirth. There will be no end to the larger-scale structures of the 3-dimensional universe because thanks to space the universe has no edge. Emphatic affirmation: there is no edge, no boundary and no exogenous shape to the Universe. We and everything else in it are irremediably thrown into life *forever* and will never absolutely die, as we evolve along a line of eternal transformation. This vision requires a re-dimensioning of our understanding of matter and life, which falls under the purview of the Quanto-Geometric transverse hierarchy in U_3. There is nothing of a referential nature in U_3 that makes one object of the hierarchy more valuable than another in its fullness. A human being is no more valuable than a star or a planet, and vice versa. However every structure positioned along the transverse hierarchy line is not only greater but phenomenologically more advanced than the previous one in the line or anyone prior. This may be construed as a result of the greater size, which means more expansive dynamics resulting in more phenomenological processes. As intelligent as a human or a *karyot* may be, he or she is not phenomenologically more advanced than a star or a planet but strictly less.

The current state of the knowledge in the discipline of Astrophysics is at the notion of 2-level-up larger scale cosmic structures above galactic structure in the 3-dimensional universe. Galaxies are currently visualized to be laid in groups, in the form of elements over the surface of vacuum bubbles comparable to foam bubbles. I must caution that these are hypotheses and that, with all due respect to our astrophysicist friends, Astrophysics has historically proven to be a tentative science with hypotheses and truths frequently proven to be wrong after the knowledge has settled.

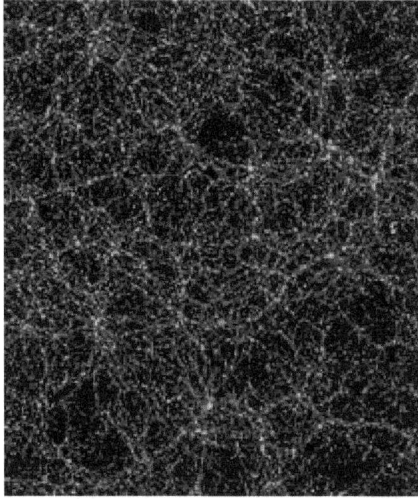

Fig. 6.3 Computer representation of larger-scale galactic groups

Consequently we are not going to engage into structural analysis of the hypothetical larger-scale galactic structures as conjectured of late. The computer model resulting from these conjectures is shown in Fig. 6.3. Given the regularity, the harmony and frankly the great structural beauty the 3-dimensional universe has accustomed us to, across all Shells up to single galaxies in terms of directly observable elements, I shall confess my refusal to believe that the larger-scale structures could be so messy, so chaotic and so ugly!

We do not at all reject the possible existence of these structures because they respond to a consequence of the Quanto-Geometric model. The reader will remember that we have conferred to the Grand Mappa Mundi Function the value of formal functional representation of the Quanto-Geometric number system or the higher abstraction system. We shall remember as well that in that abstraction every transverse number set is open to infinity. The numeric transverse hierarchy is captured in the functional representation by a circular arc of π length. The symmetry groups delineated over π are infinitely granular in accordance with the higher abstraction so that an infinite line of Shells may be accommodated by the transverse order thru that granularity. As to the s intervals, the whole *normal* interval bundle from 0 to 4σ may be re-distributed as unity over the semi-circle with a particular value assigned to each bundle, thereby making the whole functional representation still hold together. In so doing the model would account for any extension beyond the galactic Shell as their reality is being firmly established. However, we believe there is better and more fruitful adjudication of mind power and material resources in researching the implications and **applications** of the cosmological constants we have hereby unveiled instead of engaging in this breathless race after infinity. We should identify the task at hand as the amplification of the living medium for mankind both into the Solar and Galactic Shells, for the cause of advancement of spirits and means as to ourselves, and as stewards of Creation for higher purpose. Some might prefer to hear me say: the antigravity craft should be henceforth our true focus in Physics and Astrophysics, and so I say unreservedly and unapologetically. Perhaps the detection of 4-dimensional and 5-dimensional reality as well in an effort to harness vertical hierarchy.

In summary, by any account rendered from a space-dominant vantage point, an epitomization of the state of the 3-dimensional universe, let alone the Grander Mappa Mundi Universe or the Omniverse as we dubbed it, should irrevocably manifest that it is: *birthless, ageless, timeless, shapeless, edgeless, boundless, infinite and eternal!*

6.33. Apologies But No Billion-Year-Stage Evolutionary Mutations

Despite the continued success in the field of Biology in relation to the decodification of the karyotic genome, the evolutionary model traditionally upheld in that discipline is, in our view, in dire need of correction. So is the visualization of the essentials that make up the living.

First of all, the million- or billion-year timeline along which the model has anchored evolution of species as the propitiator and catalyst of mutational changes is flatly off the mark. In Quanto-Geometric analysis, within the Karyot Shell there is no time variable to effect events or any constitutional element, nor is there any in the Astral Shell, the immediate medium for Karyots. The time element must summarily be stricken out of the equation.

Secondly, the fact that a karyotic species may tangentially evolve into another does not mean that at the higher levels of the karyotic line intelligence may not re-create lower-level karyots in minute quantities or massively.

Third, the express rejection by the field of possible mechanisms by which karyotic genome may sense environmental conditions and orchestrate adequate response is highly questionable. Effectively it is very hard to object to certain phylogenetical artifacts such as mimicry whereby the genome strikingly replicates the exact appearance of the environment for distinct fitting purposes, moreover when the contention is that **billions of contentious years** of fortuitous mutational trials are necessary for such an outcome.

Fourth, the stipulation that mutations are fortuitous events, which amounts to a stipulation of the resulting traits or characters as *untyped*, is unsustainable within the Quanto-Geometric framework. There is nothing in existence, whether it is an abstract or an object, that is not typed along the line of the 9 Primitives and/or subject to intervention of the Tri-Valent Operators. Everything belongs to an order of Quanto-Geometric types. Those types are the deepest level of codification within the karyotic genome, I might add. I shall also point out in passing that here essentially lies the reason why the genome project, despite its large success, surprisingly enough found a lot less gene sequences than expected in the human genome.

Fifth, the field has been developing in the lack of the most significant defining element for an understanding of what the living really is, which is the notion of a *karyotic physical constant*. Human karyots are not the best and brightest of creation and do not enjoy something unique to only themselves called life and intelligence as opposed to inanity for the planets, stars, moons, etc. The physical structure of karyots is anchored on one physical constant to the same extent as all the elements germane to the other Shells (Fig. 6.4). The space wave function operating in karyotic cells has a much lesser level of multiplicity and auto-dynamism than those attributed to elements in the higher Shells. Thrown into ecliptic space, a human as is will shortly die. Thrown into ecliptic space, a planet like mother Earth will live and thrive. Because a planet follows an inalterable orbit around a Sun Star does not mean that it is bereft of higher level phenotypical processes, part of which may and should be construed as intelligence.

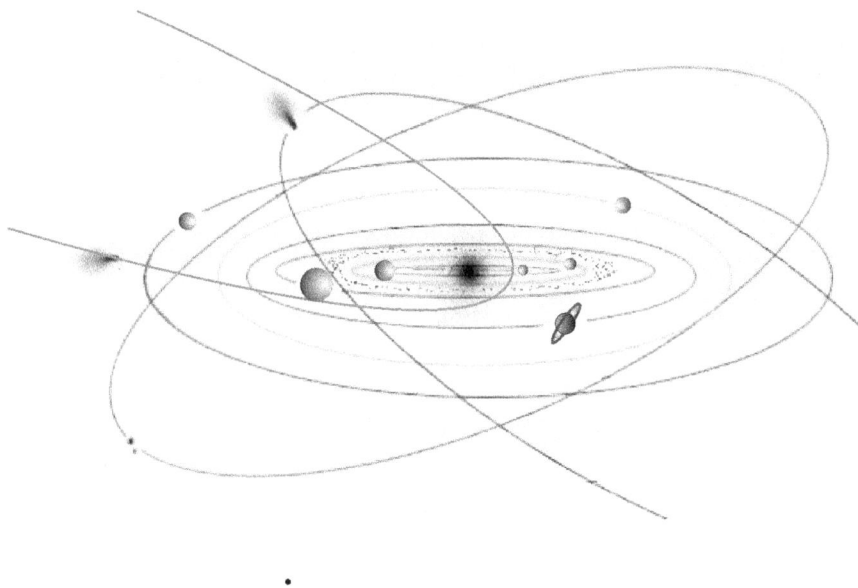

Fig. 6.4 Our solar system with orbital paths of comets and planets

Involuntary motion is inherent to the life of higher Shell beings just as it is to the life of all human karyots. The women among us duly know what their uterus cause to their spirits and body on its own in prelude to delivery and during delivery. Furthermore, does the whole delivery event itself turn a woman into a "cyclone", a "hurricane" or a "tornado"? Oh, they never stop being intelligent all the while! Just that they don't stop hollering and drifting like gusty winds and crying like rain showers.

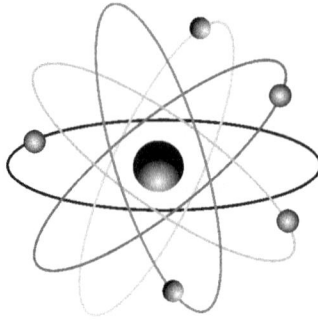

Fig. 5.5 Common Model of a multi-electronic atom

Few if any man have ever imagined their entire life as nothing more than orbiting about their homes or "homing", just as planets or comets do around the Sun Star. We wake up and go away from our homes, in order to create entropy in the world in the form of our daily work routine, doing a lot of intelligent things, and then come back home, and then go away again, perhaps somewhere else, shopping for instance, and come back home, and perhaps sleep, and go away again, and on and on again... If you should now take a look at the orbitals of an electron around an atomic nucleus as illustrated in our physics and chemistry textbooks (Fig. 5.5), or the orbiting of comets around the sun for instance, in light of the above, you now know what you have been doing about your *home* all your life and what your life was mostly about. The great things that you might think about yourself and the great things that you have done for yourself, have all been accomplished over this orbital substrate. The storyline of this orbital substrate constitutes a very large part of the story of our lives. And what conditions those orbitals for all of us is the value of our $[dQ(s)/ds] \times k_p$ function for our s value of $\sigma(8/5)^{1/2}$ as incarnated within our bodies, cells and organs. Now since the *home* is likely to shelter more than one occupant, i.e. parents, children, cousins, nieces and perhaps a friend, the orbital structure becomes more akin to that of a multi-electronic atom (Fig. 6.6) or that of our familiar Solar System. Let me close these remarks by stating once again that there is no flat object anywhere, every Shell object in its fullness is a being, and the beings we call planets and the Sun Star, of which we know so very little, are far more advanced and complex than we are and will ever be as karyots.

Last but not at all the least, human life orbitals are thoroughly carried out thru involuntary motion. Our daily circling and wandering in space requires the automatism of our legs walking in step with one another, the automatic balancing of our body thru coordination of leg and arm motion when we run, the automatism of tongue and jaws when we speak, and so on and so forth. Even when we employ an artificial craft for our daily space travel a large measure of automatism in steering the craft is a requirement for success in the undertaking. We never have to think about the minute details of coordinating our limbs in the performing of any of those *movements*, while our minds generally wander about other concerns. As soon as we start investing voluntary thoughts into those automatisms, they instantly break down. All higher-level processes are overlaid on these orbitals as epi-phenomena and require the innate automatism of the orbital substrate for their development. So it is for all karyots as it is for all beings in the higher Shells.

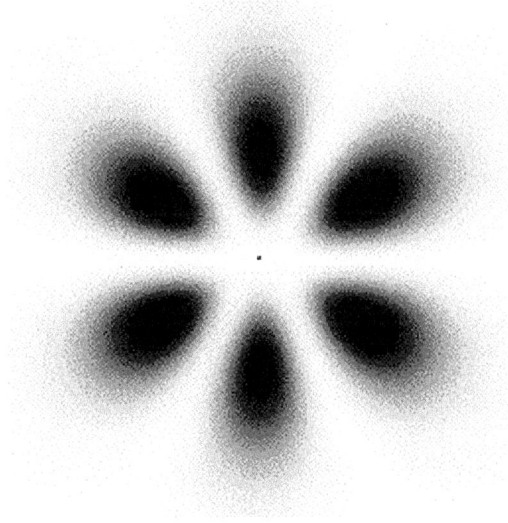

Fig. 6.6 A *4f* atomic orbital with nucleus at center in a planar view

For the readers who may remain unconvinced to this point and view this construction as an extrapolation, I shall invite them to another discussion staged as follows. Consider one of those virii whose whole body is made up of 3 juxtaposed **crystal** blocks with a **karyotic** nucleus in one of them (Fig. 6.7), an element (the virus) which is able to remain dormant for thousands of our years and become active again under favorable conditions. The question asked is three-fold:

- What of It was life when the state of inane matter was in effect?

- What of He will subsequently make him inane matter when the state of life is in effect, in the understanding that inanity can recur thereafter?

- Can He die? If so what does that mean?

Bonus question: If He may die, should He be buried?

Fig. 6.7 One of the many human-ladden virii

Finally, the above considerations turn moot the otherwise pungent question between Creation and Evolution that has long created sharp divisions everywhere in culture, whether in academia or the larger society. If indeed the precursor of the genomic molecule was amino-acids formed in Earth's atmosphere at the favor of light rays of thunder (today conjectured of comet origin), we shall conclude that a combination of cosmic intervention between *mother* Earth and the *king* Star as a process created karyotic life. In the context of those larger beings taken appropriately for who and what they are, the terms creation and evolutionary emergence become essentially exactly identical or interchangeable.

6.34. Response to Today's Challenging Action Items in Physics

There is a rather long list of challenging action items in the field pressingly awaiting unequivocal answers. We are going to review some of them in the light of the contributions the Quanto-Geometric Theory has offered thus far for their resolution. It is natural that the call for resolution of these conundrums formulates them from the standpoint of the current perspective the field holds on those questions. However, no one should be surprised that these answers may require, as they indeed historically have, a thoroughly different and wholly new perspective for the unraveling of those puzzles. The continued history of the discipline of physics has proven no less.

6.35. Dimensionless Physical Constants

The maturity attained in the field has made evident to the Standard Bodies the need for moving away from dimensional physical constants, well understood as it is that dimensional units are human constructs and not attributes of matter. The problem, however, with the few dimensionless physical constants that are known is that they cannot be independently calculated or formulated thus far but only estimated from other experimentally measured physical quantities by applying relationship formulae. This has remained a major unresolved problem of physics until Quanto-Geometry has knocked at the door. Effectively we have presented here a theoretical system able enough to predict not one or two physical constants but a whole host of constants in a dimensionless manner. When one rationalizes these known constants by expressing them in a single system of units and making sure they are consistently computed from formulae and relationships in compatible units, one finds complete agreement between the known values of these physical constants and the Quanto-Geometrically derived ones. The non-physicist reader might have indulged in some objection to the fact that the known quantities per their dimensional units are not quite represented in the Quanto-Geometric figures. Matter-of-factly this is exactly what they cannot and, most importantly, **are not expected to be by the field**, and it is a rather remarkable fact that despite the absence of unit dimensionality in the Quanto-Geometric figures, they each bring a figurate match to their par on the table that is frankly no less than astonishing. Not having had anything to do at all with the dimensional units, and given that these units are man-made constructions, one might have rightfully expected an independent theoretical system to cast completely different numbers or quantities. By not doing so and showing up as matching figurates, the Quanto-Geometric figures curtail any argument about identity and correspondence, or lack thereof, with the known dimensional quantities. Otherwise we would need to engage in a discussion about correspondence and compatibility between the physical constructs and laws behind the known constants and the constructs and abstractions behind the independent system. Therefore, with these results, the perspective and axiomatic behind the new theoretical system ought to bear preponderance over the former, while the most intractable visualization Quanto-Geometry, as the new system, uncompromisingly upholds is the absolute death of time, henceforth passing the weight of all evidence to space. Further, there is no doubt that Quanto-Geometry represents the mathematical physics axiomatic system envisioned by topmost 20th Century mathematician David Hilbert in his famous 6th Problem.

6.36. Do Physical Constants Remain Constant Over Time?

Although this interrogation proves to be frankly of no practical significance, despite its hypothesized connection with the alleged cosmological constant for an expanding universe, it must be demoted as a concern because it expresses no more than a moot problem. Emphatic answer: No, physical constants do not change in time because there is no universal timeline for them to change along.

If we move past defunct time however, worthy of attention is the fact that the Quanto-Geometric physical constants are rooted on σ as a parameter: their values exclusively depend on the standard deviation of the Grand Universe which includes U_3. Because the σ parameter may theoretically take a number of different values, the correct answer to the question would seem to be: yes, the values of physical constants may change.

However, the Quanto-Geometric function is not the proper framework for the treatment of this question. This question shall be analyzed within the framework of the Quanto-Geometric numeric abstraction rather because this abstraction is what primarily governs physical reality in accordance with the postulate that physical reality stands for applications of the higher numeric system. Consequently the question of the invariance or not of physical constants remits to the consideration of whether or not the *normal* stack of number systems may inherently change. From that standpoint, the categorical answer to the question posed is that **physical constants do not vary**. The reasons for the negative are three:

1. Despite their open-ended trending to infinity, the cardinality of each number set in the stack is higher and remain higher than that of the previous.

2. Their individual cardinality must remain proportional to one another, lest the whole system is disjunctive.

3. We do not see any meta-level of abstraction that would pose, cause or justify an intrinsic asymmetric change in the modulus of sets' cardinality across the stack. Any change, if ever, would have to be symmetrically applied across the stack, leaving the standard deviation of the entire stack unchanged.

Consequently physical reality as a projection of the *normal* numeric system must have a fixed standard deviation figure. Hence the **invariance** of physical constants.

The fact that the actual known constants, attained thru a completely foreign system, present figurate numbers with a high match with the ones derived from functional abstraction seem to be a testament to the above conclusion. We put the visible mismatch beyond the first 4 or 5 significant digits on account of inaccuracies arising from improprieties of our unit system per its theoretical construction as well as the measurements themselves.

Finally it is worth underscoring that experimental results corroborate our theoretically-driven conclusion of the constancy of physical constants. In effect, in December 2013, physicists studying methanol molecules in a distant galaxy reported that the variation of proton-to-electron mass ratio μ in those molecules measured a $\Delta\mu/\mu$ change ratio equal to 0.0×10^{-7} at red shift $z = 0.89$, consistent with a "null result" in their own terms.

6.37. Fundamentalism in Physical Constants

The reader may have noticed that, in constructing the Quanto-Geometric Theory, we have avoided as much as possible employing the descriptive epithet of "fundamental", an overused term in most literary cultures under the Sun. By that trend, everything and anything of some significance or importance in the global culture emphatically becomes "fundamental" whether in writing or in language. We have shown that in the highest forms of abstraction there is a whole order of things *fundamental*, the term quickly becoming too poor to establish the necessary distinction between one construct and another.

In the spirit of the primacy of dimensionless physical constants, physicists, thru regulatory bodies of the field, have elected to raise some physical constants to the status of *"fundamental physical constants"* and have chosen the dimensionless constants to play that role. The term "fundamental physical constants" however has also been used for such constants as the speed of light, vacuum permittivity, the gravitational constant and Planck constant with regards to their universality and large scope of application, this despite their dimensional nature. That seems to stand as a regrettable but apparently inevitable confusion. The cause of this situation seemingly stems from the field not perceiving admittedly the higher order from which these quantities arise. Quanto-Geometry has now opened the door to that order by showing how a Tri-Valent Operator has formulated the hierarchy of those constants. If any physical constant shall be deemed fundamental, it should be the speed of light constant to deserve that role, in fact the Quanto-Geometric functional figure that directly relates to it rather, which is the value of the Grand Mappa Mundi function for the 3-dimensional universe. The black-body-related constants should play the role of secondary physical constants whereas the interaction coupling constants defining the U_3 Shells, the known ones among them coined fudge factors, should play the role of tertiary physical constants. The next level down of constants should be viewed or ranked as epi-phenomena, as important as they might be in the configuration of matter. As to the mass constants attributable to sub-component particles, they should be regarded as infrastructural constants.

Some physicists have made laudable efforts in framing the questions relating to fundamental constants in terms of what in their judgement ought to be found regarding these constants going forward. Martin Rees's Six Numbers proposal is probably one of the most prominent of these sets of deliberations. I will let readers exercise their own appreciation on these deliberations and re-dimension them within the Quanto-Geometric context by themselves.

In the aggregate, we are in a position to conclude that the long-sought goal in theoretical physics of finding an abstraction that organically yields all the fundamental dimensionless physical constants and successfully weighing them against measured values are hereby significantly met. The remaining constants that address the mass of elementary particles we will address in the next cast of this study. By and large, we now may have a wholly stratified *chart* of physical constants because we know the relationship they keep or must keep with one another as much as the hierarchical order of their distribution.

The elusive goal of this Theory of Everything able to perform that far-reaching calculation has been significantly met with Quanto-Geometry at this stage.

6.38. Unification of Physical Units of Measurement

The Quanto-Geometric framework not only adequately orchestrates the unification of physical forces but, interestingly enough, it also unifies the scaling nomenclature for units of measurement, which are otherwise subject to man-made circumstances and representations, hence to a string of inevitable inaccuracies. In effect, the majority of measurement units in use in the physical sciences bear in one way or another the name of their inventors in honor to the scientists who discovered the physical quantities or phenomena they are associated with. While this practice remains perhaps of no harmful consequence, it sure does nothing to alleviate the necessity to keep and secure samples or etalons of matter in special conditions able to guard against the slightest variations. Since those conditions are not easily achieved, the Standard Bodies have decided to move away from the practice of keeping etalons by increasingly electing to peg the value of dimensional physical constants to that of the dimensionless ones, as previously mentioned. The risk in so doing is that the imprecision in one unit of measure may quickly become compounded, adding even more inaccuracies to the unit system as a whole.

By contrast, the Quanto-Geometrically derived constants are all parameterized on σ, the standard deviation figure of the Omniverse, which appears as their natural unit of expression. The many different mathematical dimensions of σ produce the specific units that measure the associated generic physical element, force or physical phenomenon. Table 5.3 reports the following dimensions of σ: σ, σ^{-1}, σ^2, σ^{-2}, σ^3, σ^{-3} and σ^6. The sweeping advantage of this system is that it provides a single minimal measurement nomenclature applicable at all scales of matter from the Sub-Particle Shell to the Galactic Shell, and even across the vertical hierarchy of the Omniverse, the meta-physical World planes. This system would dramatically reduce the necessary number of etalons to one, and only one, simplifying the institutional task of standardizing and safeguarding physical units of measure and their etalons.

To anyone harboring some reservation about the reported physical constant quantities in this study toward the above proposal, I shall further offer the following concession. In the absence of a protocol or standardized estimation method that sets the acceptable level of precision of any quantity or constant theoretically derived, those reservations are perfectly understandable and probably justified. The problem however is that there exists no referential framework against which to contrast those derived quantities. The historically known units of measure do not hold any referential value, unless we expressly choose to establish them as such, with all their known inaccuracies, for the sake of relying on at least one independent system. We recognize this all amounts to a very fine needle to thread but it is an inevitable task for the new Physics.

Furthermore, in assessing the authenticity and accuracy of the Quanto-Geometrically derived figures presented in our chart, one possible method for the estimator is to start out with the best match, opting in doing so to set out with the highest level of certainty. One then analyzes the relationship between this quantity and the next one in its class as well as the top one it is directly derived from. One makes a critical judgement about the level of accuracy of those dependencies and the soundness of these relationships in theory. The estimator walks throughout the chart in that manner until c, the primordial constant for U_3, is reached. This exercise, though not sufficiently systematic, may nevertheless provide a satisfactory basis for assessing the validity of these results in the interim.

6.39. The Standard Model and the New Physics

Despite the harsh criticism extended here in regards to several of its concepts, the Standard Model is a Theory that has probably correctly served its original purpose. Essentially of taxonomic nature, it has always been implicitly intended, however, to be a provisional theory, with the mission to adequately shelter and assimilate new experimental findings occurring in particle physics specifically. The intention behind our criticism is solely but uncompromisingly to obliterate the notion of time anywhere it might appear in theory. It is indeed a tall order for the Standard Model to continually stimulate and harmoniously integrate new discoveries, however challenging they might turn out to be, to its core principles. Not to mention the positive support role played by the Standard Model in the effort to standardize and institutionalize modern physics, in particular when it comes to experimentation in high energy physics. In that sense, this Theory has served us well since its inception.

Furthermore, the multiple generations of physicists who have strived under that banner, many of profound intellectual brilliance, but all of whom with an unwavering determination in the quest for unraveling the mysteries of matter, deserve no less than earnest praise for their efforts. So do String, Loop Quantum Theoreticians, and others of other color, who have been engaged in the exceedingly arduous task of building an overarching wrapper of quantum-gravity unification over the Standard Model in order to overcome its lacks and shortcomings. In the community of physicists at large, there has been a long string of physicists who have demonstrated the commendable ability to join personal valor and selflessness to intellectual brilliance at crucial junctures of the history of mankind. The world has the community of physicists to tribute for the Industrial Revolution, the cataclysmic deterrent stamp to the official Nazzi-Japanese-Fascist global project of folly following its defeat (regrettable but arguably necessary), human space travel to the Moon, the global Information Technology infrastructure, to name a few of the most salient achievements.

Without a doubt, physics is and will continue to be a paramount force in human culture. Yet, with all the deserved credit granted to the Standard Model, and in light of enormous challenges currently facing the Theory as we shall see in Volume II, it is now the juncture for a new vision to take the stage.

Chapter 7

*To see what is in front of one's nose
needs a constant struggle.*
George Orwell

QUANTUM PHENOMENA
IN THE MACROCOSM

As we mentioned before, the debate of all debates in modern physics has long been the question concerning the completeness of Quantum Mechanics, a controversy owing to the mathematical treatment of statistical nature at its foundation. Because probability or statistical theory is overwhelmingly a numeric science with no contemplation or inference with regards to causation and determinants of action, physicists have always felt that the Quantum theory is at a miss when it comes to intimately intelligible interpretation of dynamic systems. Even the proponents from the Quantum School of Copenhagen felt that a philosophical justification was necessary in order to at least provide a view as to the analytic implications of the Quantum Theory, notwithstanding the pressure brought to bear onto them by compelling critics. Amidst this hurly-burly, little to no attention was ever paid to the dimension of discrete Quantum expression in the macrocosm thru the application of the same science of Probability Theory. In the following, we re-examine Probability Theory to show that it encloses a discrete ontological dimension that is universally applicable to all objects permeable to probabilistic treatment, thereby directly relating macroscopic dynamics to quantum dynamics on the basis of a unique Unification theory.

7.1. Applicability of the Grand Eigenfunction at Large Scale

The question of unification of Quantum Mechanics to General Relativity is most easily resolved by considering how quantum phenomenology is rendered in the macrocosm under the auspices of the Quanto-Geometric Grand Eigenfunction and its derivates. The Quanto-Geometric canon, as a symmetry stack bundled in the Grand Eigenfunction, describes the ontological foundations of the continuum of coordinate space and the discretisation of the quantum-scalar, along with their correlations. These ontologies are implicated in all objects and phenomena permeable to statistical analysis thru the Gaussian distribution, which is a derivate of the Grand Eigenfunction, per incipient analysis in Chapter 2. In that sense everything in our immediate or remote living environment that is permeable to the standard Gaussian distribution constitutes evidence of quantum phenomenology right in the midst of the macrocosm.

7.2. Timelessness in Statistical Analysis

The most notable yet least appreciated result in statistical analysis is the absence of the time variable in standard distributions, for all manner of phenomena, all of which are initially and ordinarily inconceivable to us if removed from the trace of the arrow of flowing time. Effectively, the distribution of heights in a population, to take just one example, is an artifact that is totally independent from the passage of time as we understand it. There is no formal statistical time constraint imposed on sampling in order to test for the standard distribution of this parameter in a population. Populations of humans throughout the ages have developed consistent with this configuration pattern for the inception of individual heights. The correlation of events and all manner of genetic and behavioral associations between individuals that will occur within the sample group, and which one may want to subject to time flow at least descriptively, has no bearing whatsoever on the distribution.

The timelessness artifact implicated in statistical treatment has been underappreciated throughout the many developments in statistical physics, especially in its application to the physics of the macrocosm. One may even further argue that one of the reasons why probability theory fits so well to the resolution of E. Schrodinger's equation independent of time culminating in the quantum Hamiltonian is because of its complete estrangement with the time variable.

7.3. Intricacies of the Gaussian Probability Function

The reader who has previously been exposed to the Quanto-Geometric Eigenfunction may have at first rebuked it charging that it is simply a repackage of the long-known Gaussian Distribution function. However several characters of that function provide indices for unsuspected features of the Gaussian that are worth analyzing in the context of this re-examination.

From conversion of the Grand Eigenfunction to its implicit form as undertaken in Chapter 2 Section 6, we obtained at one step:

$$\frac{-s^2q^2}{2q} \cdot \frac{2}{q\sigma^2[2\ln q + 2\ln \sigma + \ln(2\pi)]} = 1$$

One may quickly notice that if s takes a zero value, then the absurd result of $0 = 1$ ensues. The variable q cannot take a zero value, since $ln(0)$ is undefined or forbidden leaving the entire left term undefined in that event. Therefore implicit in the function is the condition that neither s nor q can be equal to zero. These constraints are of paramount significance when it comes to interpretative treatment allowed by the Function. In the explicit expression was already set the condition that the Function may never take a zero value $(q = 0)$ because we could already note that as s becomes infinitely large the function approaches zero value without ever taking that value, making the s-axis equivalent to an asymptote. The implicit expression thus reinforces the constraint on q at the same time that it dictates an equiparable constraint on s $(s \neq 0)$.

The constraint of no q-intercept is not known in Statistical Theory. However, it bears significant phenomenological consequences in analysis inspired by the Function. Should we take q for representing probability density and if s models measurement spread for instance, the constraint of no q-value for s becoming extremely small remits to two significant conclusions:

- No matter how many iterations in measurement in the pursuit of precision, there will always be a probability gradient of error, as small as it might be. In other words absolute certainty (100%) is not possible.

- There will always be a minimal probability gradient of precision, no matter how removed or inaccurate the measurement. In other words, zero certainty, which is equivalent to absolute uncertainty, is not possible.

Very particularly, because the Function admits no q-intercept by reason of discontinuity at $s = 0$, the notion of total area under the curve and that of said area equaling unity completely vanish. The notion of the area under the curve equaling unity, however, stands for the most fundamental axiom of traditional Statistical Theory. Since the function is simply not smooth, thus not integrable, over the full domain of $[-\infty, +\infty]$, the total area under the curve of the Quanto-Geometric function cannot be formulated.

7.4. Discretisation in Probability Theory as a Poster Child of Quanto-Geometric Theory

In quanto-geometric analysis, q or $Q(s)$ represents the universal scalar which ontologically gives rises to mass, and s represents universal space or the void spread. The above characterization poses the first restraint on both the continuum of space in any envisaged dimension and the discretisation of the scalar wherever manifest.

Therefore, if measurements of any kind will always incorporate a gradient of uncertainty or error, if data of all kinds gathered in macroscopic phenomenology will always miss a certain mark, it is not just a fact of nature without cause. They are discretisation restraints imposed by the foremost form of distribution at the root of all ontologies and phenomenologies.

7.5. The Problem of Meaning and Causation in Statistical Analysis

While Statistical Theory have been widely used historically throughout the sciences, it is fair to say that it has always left scientists with the unsatisfactory after-taste of incomplete grasp of the actual physical dynamics effectively correlating the variables at play. Statistical accounts do not explain the "why" it happened and only gives a countable account of events, devoid of underlying dynamical meaning.

A large number of groups in the physics of matter at the macroscopic level and the life sciences follow the distribution pattern depicted by the Probability Density Function. Its use is indeed very wide spread in science in general. Below we report just a few of the well-known areas where prospected data is known to conform to the Gaussian distribution:

- Individuals' heights in populations of the living

- Blood pressure in human populations

- Measurement errors in science

- Test grades in education

- IQ scores in psychometrics

- Workplace salaries

- Societal Polling Data

While statisticians do not ask more of the capabilities of this Function and the empirical analytic techniques developed around that Function, in the physical sciences and in particular in certain quarters of theoretical physics, statistical analysis of dynamic systems had historically been considered incomplete, leaving analysts in search for deeper meaning. The nature of statistical physics had been at the center of the famous debate over

Quantum Mechanics ever since the publication of the EPR paper. Effectively causation and determinism are completely absent from this form of knowledge. Most feel that the sole rendition of a probabilistic percentile of events does not constitute thorough explanation.

7.6. Standard Deviation Metric

The ignorance of the inner meanings or ultra-structure of the variables of distribution has led to empirical techniques put in place in statistical theory, thereby obscuring the higher order of certainties inherent to macroscopic physical dynamics.

The standard deviation value represents the most critical parameter set forth in statistical theory for measurement metrics. The empirical practice has it that the abscissa of the putative coordinate system for the graph is measured in integer multiples of σ, setting the basis for the so-called empirical rule of 99-95-68. There is nothing in nature or inherent to the theory, however, that confers any particular weight to these percentiles. According to this rule:

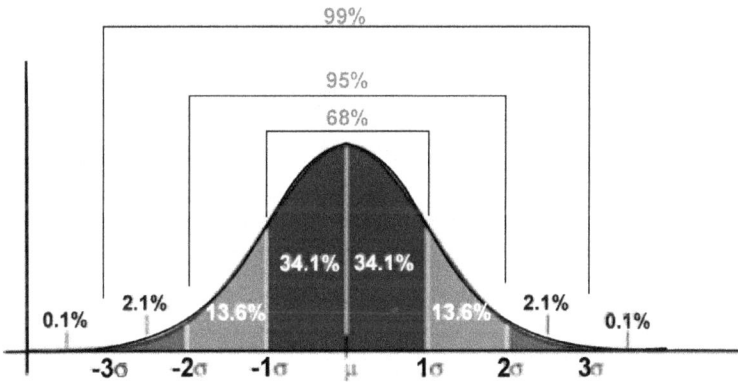

Fig. 7.1 The empirical rule of percentile distribution in Probability Theory

99% of the data corresponds to the area under the curve delimited by -3σ and $+3\sigma$.
95% of the data corresponds to the area under the curve delimited by -2σ and $+2\sigma$.
68% of the data corresponds to the area under the curve delimited by -1σ and $+1\sigma$.

An empirical σ metric had been devised for a seemingly rational treatment of sampled data. The one particularity about the standard deviation point that grants it a manner of referential weight is its singular inflexion gradient relating to the curvature of the curvilinear pattern of the function, reflected in its unit value of the abscissa scale.

7.7. Overwhelming Numeric Trend in Statistical Theory

The emphasis on number play is such in Statistical Theory that it might almost be considered a form of Number Theory in its own right. It is inarguably a science more heavily based on querying probability figures than on studying *behavior* per se or phenomenology if at all. Accordingly, the queries are as follows.

A dice is rolled. What is the probability for an even number among the six to show when it stops?

If the probability for the sought event is *p(E)*, *n(E)* the total count of the sought event and *n(S)* the total number of possible events, the probability of the sought event is given by the formula:

$$p(E) = \frac{n(E)}{n(S)}$$

In this case:

$$n(E) = \{2,4,6\} = 3$$

$$n(S) = \{1,2,3,4,5,6\} = 6$$

Therefore:

$$p(E) = \frac{n(E)}{n(S)} = \frac{3}{6} = \frac{1}{2}$$

$$p(E) = 0.5 = 50\%$$

The above allows us to coin the more general formula:

$$Probability \ of \ Specific \ Events \ = \frac{Number \ of \ Specific \ Events}{Number \ of \ Possible \ Events}$$

A probability is thus a figure that quantifies the specific with respect to the general, at the same time that it principally concerns physical *events*. Furthermore all probability of specific events must be less than or equal to 1. So it is simply because the number of specific events constitutes a subset of the number of possible events, making thereby the fraction always less than or equal to 1. Additionally, because the probability is a ratio figure and that division by zero is forbidden, the probability cannot be equal to 0. Which leads us to the following conditional statement:

$$0 < p(E) \leq 1,$$ where $p(E)$ is the probability of specific events.

The pioneers of statistical theory (Poisson, De Moivre, Lebesgue, etc) have found that the Gaussian distribution function perfectly encapsulates all of these principles and have erected this function as the pivotal mathematical scheme for the study of facts or events of chance. Statistical theory views a statistical fact or event as a simple mathematical entity whose only virtue is to contribute to a probability figure. It is not concerned, nor does it consider that it should be concerned, with the underlying physics implicated in the event.

Once the statistical principles have been transposed onto the Gaussian distribution function, statisticians have built an interpretative structure allowing for the computation of probability percentile figures from the data that might be empirically available from a population under scrutiny. Particularly useful to that aim is the z-score formula:

$$z = \frac{X - \mu}{\sigma},$$ where μ is the mean, σ the standard deviation and X the variable's space

separation being studied.

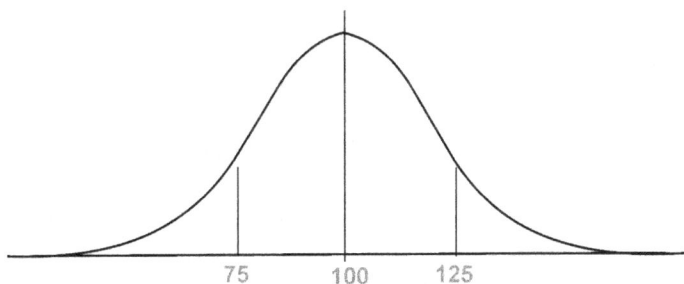

Fig. 7.2 In Probability Theory the graph of the Gaussian distribution
is not based on a Cartesian coordinate system

Note that the treatment of the abscissa incarnating the space variable is somewhat unorthodox insofar as the Cartesian coordinate system, since in the Cartesian system the mean is always 0, mid-point between the set of negative real numbers and that of positive real, in accordance with the real number line. In that sense, the abscissa on the probability coordinate system is only a surrogate of the Cartesian abscissa (Fig 7.2).

From the z-tables, to the standard deviation figure, to the μ figure, statisticians have developed empirical techniques aiding in the computation of probable percentiles. While this is all good and well from a utilitarian purpose, the question concerning the meaning of the underlying physics of statistical events has never been addressed, not even in mathematical-physics. The problems posed tend to become merely numeric in nature, such as illustrated below for the use of the empirical formula:

$$s = \sqrt{\frac{\sum (x - \bar{x})^2}{n-1}}$$

(1), so-called unbiased version or its equivalent:

$$s = \sqrt{\frac{\sum (X - \mu)^2}{n}}$$

(2), so-called biased version.

A typical reductionist problem follows:

Find the standard deviation for the following distribution of numbers: 13, 16, 18, 21, 31, 44, 45 and 55.

Step 1: Using expression *(1)* and adding up the 9 numbers in the sample or data set:

$x = 13 + 16 + 18 + 21 + 31 + 44 + 45 + 55 = 243$.

Step 2: $\dfrac{x^2}{n} = \dfrac{243^2}{9} = \dfrac{59049}{9} = 6561$

Step 3: Squaring and adding up the set of original numbers:

$\bar{x}^2 = 13^2 + 16^2 + 18^2 + 21^2 + 31^2 + 44^2 + 45^2 + 55^2$

$\bar{x}^2 = 9137$

Step 4: Subtracting the amount in Step 2 from the amount in Step 3 yields:

$\dfrac{x^2}{n} - \bar{x}^2 = 9260 - 6561 = 2576$

Step 5: We subtract 1 from the number of items in the data set, because variance estimated on the basis of n-1 is said to be unbiased:

$9 - 1 = 8$

Step 6: By dividing the number in Step 4 by 8, the resulting number in Step 5, we obtain the variance:

$2576 / 8 = 322$

Step 7: The square root of the variance represents s the standard deviation figure:

$$s = \sqrt{219.778}$$

$$s = 17.94$$

The above example illustrates the pervasiveness of pure numeric treatment commonly undertaken in statistical analysis.

7.8. Numeric Focus in Probability Theory

Two central theorems in Probability Theory significantly contribute to the numeric focus of the science, notwithstanding their unquestionable accuracy. One is the Law of Large Numbers, and the other the Central Limit Theorem, both intimately related to one another.

7.8.1 Law of Large Numbers

The Law of Large Numbers expresses the fact that the numeric characteristics of a random sample become closer to the characteristics of the whole population as the cardinal size of the sample increases.

For example, a single roll of a fair six-sided dice produces one of the numbers 1, 2, 3, 4, 5, or 6, each with equal probability. Therefore, the expected or **mean** value of a single dice roll is

$$\mu = \frac{1+2+3+4+5+6}{6} = 3.5$$

Consonant with the Law of Large Numbers, if the number of rolls of the six-sided dice becomes large, the average value of the total sum of the top faces shown or the mean value is likely to be close to 3.5, while the precision increases as the number of throws or trials increases. By the same token, this law establishes a correlation between the variable space and the abscissa of the coordinate system underlying the Gaussian Distribution Function. Because each side of the dice has a probability of 1 or 100% to appear when the dice stops, there is correspondence between the probability value of 1 or 100% with the μ mean value for total number of trials: the mean value must coincide with the probability *density* axis.

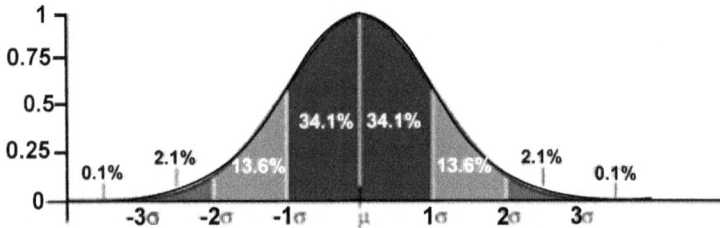

Fig. 7.3 Despite acknowledgment of a vertical coordinate axis by the Law of Large Numbers, the axis is not officially acknowledged in Probability Theory

In the empirical system of Probability Theory however, the vertical axis of the coordinate system is uncharacteristically not acknowledged, because the system technically privileges the area under the curve in the determination of probability values.

7.8.2 Central Limit Theorem

The Central Limit Theorem establishes that, for the most commonly studied scenarios, when independent random variables are added, their sum tends toward a *normal* distribution even if the original variables themselves are not *normally* distributed. The foremost normal distribution is the commonly known Gaussian distribution. This theorem generalizes the law of distribution on account of trial counts as it stipulates that all distributions, no matter their original forms, tend to conform to the normal distribution as the trial counts reach a larger and larger number. On that account, probabilistic and statistical methods become universally applicable since they can be applied to many problems involving other types of distributions.

Just as the Law of Large Numbers principally correlates to the horizontal axis of the probability coordinate system, the Central Limit Theorem directly correlates to the putative vertical axis of the probability coordinate system in that it establishes the same to be the materialization of the scalar state of a normal distribution. The larger the scalar state of the sample, one should conclude, the more resolved the normal distribution. In that sense, what we normally call the probability density as it relates to the vertical axis of the Gaussian distribution function, must in fact be equated to a scalar density.

7.9. Toward the Physical Meaning Underlying Statistical Accounts

We are next going to undertake an inquest of the Gaussian distribution with the aims of uncovering analytic meaning beyond pure numeric accounts.

7.9.1 Re-Dimensioning the Standard Deviation Metric

Following the vein of predominance of the role played by σ the standard deviation figure, one should pay attention to curvature evolution throughout the rundown of the function. This undertaking reveals a different metric to be superimposed on the abscissa of the coordinate system. That is the metric arising from the curvature sectionals making up the curvilinear pattern of the function. A study of these sectionals in terms of Tensors and Operators has been undertaken in Chapter 4. Most importantly, the symmetry implications of this approach contribute to the most fundamental order of certainties ever revealed to be associated with the Gaussian distribution as Quanto-Geometrically redimensioned. Effectively every sectional corresponds to an ontological and phenomenological Norm characterizing both the structural and evolution modalities of every object in the universe.

It follows from the above that the distribution is best normalized transversally by catering to the curvature sectionals as strata across the q-axis, given that the discontinuity at $s = 0$ further squarely forbids application of the 99-95-68 % empirical rule.

This set of Nine Norms represents the highest order of certainties encapsulated in the Gaussian distribution. It incontestably projects to every group or topology of objects fit for statistical analysis. The sigma super-metric established over the s variable (Fig. 4.4) represents the symmetry metric grading the fabric of space in its continuum, while the values of the q variable establish the true first-order quantization basis applicable to fitting topologies. Note that we are expressly avoiding the characterization of quantization of space because the space variable shall always be visualized as a continuum and not a bundle, given that what essentially distinguishes one space spread from another is not a quantity but a quality, to be precise a geometric or symmetry quality.

We hope that the reader is fully taking stock of this development which is increasingly deflating the problem of applying analytic schemes to the macrocosm that are at the same time valid for the quantum realm. I cannot overstress that to the extent that statistical theory indeed contains the complete quantization basis and the geometric and symmetry elements characterizing the continuum of space, it brings to every observable object or subject of analysis in the macroscopic realm this quantum background long coveted for a unified view of physical dynamics.

7.9.2 Event Transformations in Probability Theory

In this interpretation of Statistical Theory, what we have arrived at thus far is a three-fold conclusion:

- The Theory implies a description based on a coordinate system with a horizontal axis for true coordinate space.
- The Theory implies a coordinate system with a vertical axis for the representation of scalar density.
- Statistical events may have physical meaning beyond the probability percentiles.

We have arrived at the above three-fold conclusion despite the fact that the Theory heavily leans toward a flat numeric treatment of distributions. Taking into account the physicality of the elements participating in a distribution, we may define statistical events as follows:

An event stack (or a phenomenon) is the larger frame of spatial transformations under which a variables set is preserved as an invariant agent to the action.

Under that vision, an event stack ultimately constitutes a new object based on two distinct variables. The randomness of the numerous iterations by the agent of action creates a new scalar state that is distributive in nature. That much becomes quite clear when we consider the throw of a dice not from the point of view of the thrower but from the point of view of the dice. Independently of how long it takes to repetitively throw the dice, or how long the intervals between the trials last, the dice as the point object of action transforms into a new scalar spread, if we allow ourselves to become blind to the throwing hand and if we run the clock on the event stack. The mass of the dice acquires a new state, of a distributive nature, that pertains to a **different scalar order** than the original and still invariant-compact state.

Likewise if we give proper consideration to the position spread of the dice at every throw, it becomes manifest as an entity in its own right. Wherever the dice is being thrown, the immediate physical boundaries of the support medium constitute the *amplitude* of the position spread experienced by the dice. It must be understood as the *stationary limit* of the action which may be or must be visualized as the wavelength of a stationary wave developed by the very *position space* or position spread. This position space spread is real coordinate space in nature, not a fictitious mathematical space, the one to which the Quantum Theory has unfortunately accustomed us. The many different iterations of the throw within this position space spread creates the *harmonic character* of the action within the spread, independent of the direction of the momentum in development. From the point of view of the dice, as a point object it is subject to or develops on its own a (repetitive) harmonic motion constrained by a position space spread. We have just described nothing else than the physicality of the material *wavefunction*, as mysterious and controversial as it has always been. It is a harmonic motional artifact that springs from the very void of physical space.

The two new variables that we have uncovered thus far are ***distributive scalar state*** and ***position space spread***. They will be directly and straightforwardly modeled by a Cartesian coordinate system whereby the vertical axis incarnates the first as the dependent variable and the horizontal axis incarnates the second as the independent variable.

7.9.3 From Variance to Ontological Covariance

In the traditional probability graph, abscissa and ordinate axes are not correlated, simply due to the inexistence of the ordinate in the graph. Therefore the variance expressed in terms of factors of σ on abscissa does not have any particular meaning in reference to a putative ordinate variable:

$$\sum (X - \mu)^2 = N\sigma^2$$

Hence the variance admittedly has very little meaning and only roughly gives an idea of spread. It is knowingly not used for much at all, except to compute the standard deviation figure, which is the main protagonist of the show. However, the concept behind the variance is a remote prelude to the very important concept of ontology or physical meaning ascribed to the abscissa variable of the Gaussian Distribution Function. As a descriptor of spread, the concept of variance relates to the notion of ***position space spread*** that we spoke of previously for the abscissa of the Quanto-Geometric coordinate system. It indeed refers to the inner variability of physical space spreads which *covariantly* relates to scalar density as the second tenet of the ontology subtending all statistical physical objects, not to say universal objects.

7.9.4 Parallel Between the Quanto-Geometric Covariant Spectrum and Statistical Percentiles

Suffice it to translate standard statistical accounts to the Quanto-Geometric canon for all the blanks of statistical analysis, not just to be filled in by causation, but to be replaced by an order of normed certainties representing the highest level of knowledge universally attainable. So it is despite the fact that the metric on the abscissa variable distinctly and more stringently goes up to 4σ in Quanto-Geometry, beyond the ordinary 3σ used in empirical analysis. Table 4.1 reflects the Primitives of the Quanto-Geometric Order of Certainties.

The referred symmetry primitives constitute the analytical categories that belie the empirical z-value calculations of gradients of probability involving σ and variance ordinarily undertaken in statistical theory. As previously highlighted, there is normally no meaning other than flat numeric cardinality to a percentile value of probability. What this science had ignored, having no means to otherwise inquire, is that there exists a transverse normative order to the probability distribution which is given by a different metric of σ over the variable. This analytic framework considers not the area or fractions of area under the curve but the curvilinear sectionals of the curve. It is there that statistical results acquire ontological and phenomenological qualities making up the deepest and most fundamental descriptive order of certainties in nature. In the strictest sense, the visualization of an area under the curve that spans across the probability density axis is illegitimate because the implicit expression of the function mandates discontinuity at $s = 0$. It is no wonder that there exists no natural or intrinsic meaning to the parcelization of the area under the curve and that the founders of this science had resorted to empirical methods of computation garnered in the z-tables in order to draw meaning and usefulness to the Function.

In a translation of the old methodology and its results to the new quanto-geometric dimension of the distribution, we would have to discard all z-values implicating transverse areas under the curve crossing the probability density axis. Only the individual tail z-values (involving external arms of the distribution) would remain valid.

For example, in educational settings, the bell curve is known to perfectly categorize test results of any class. Results are always such that the bulk of students will score the average or a C grade, while a smaller number of students will score a B or a D. Yet an even smaller percentage of students will score either an F or an A. The question is what is the cause of this distribution or what does it mean?

The Quanto-Geometric Eigenfunction, grandfather of the Gaussian distribution function, teaches the following:

A.- The position space spread axis mandates a division of students in two groups (Fig. 7.4). One group that positively interacts with the study material (positive s-axis spread), making an honest and consistent effort to apprehend the material. One other group that negatively interacts with the material (negative s-axis spread), engaging inconsistently with it. Per the axis, both groups are mirror images to one another.

B.- On the positive side, the 1% of students obtaining an A grade, experience the largest *dispersion* or *detachment* in their learning experience with the material. By exerting the ability to take *cognitive distance* from the material they are able to better prospect and visualize the material. This ability is a characteristic of the Quanto-Geometric asymptotic or transcendental layer where the s value set of the wavefunction is at its maximum.

C.- At the other end of the spectrum, we have the about 30% of students, the largest group, with the average C grade. These are the students in this group who could realize the least amount of cognitive detachment from the material. Their interaction with the material was too constraining or intrusive, or perhaps obsessive; they were unable to *elevate* over it for better visualization and thus better grasp. This ability or better say *inabili-*

ty is typical of monocentric ontology experienced primarily at the monolithic layer of the Quanto-Geometric spectrum.

D.- The intermediate group of students in this rubrique, quantified in the distribution at about 13%, are those obtaining a B grade. Their B grade is due to a cognitive *distal* ability that is below the A group but above the C group. They are essentially marked by the Quadratic ontology experienced in the spectrum at level 4, where position space spread equals scalar density in covariant influence.

One can see that, by catering to the ontological variables, the distribution acquires meaning much beyond the percentile figures. If we were to be more granular about the percentiles and taking into account the full set of expected grades (A+, A-, B+, B-, etc.), and further delimit them according to the σ super-metric, the quanto-geometric set of 9 transverse primitives would become pointedly manifest in the distribution for us to observe.

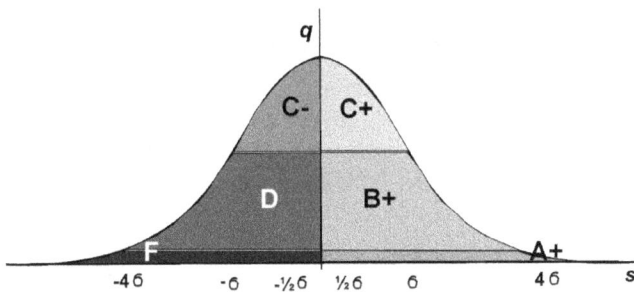

Fig. 7.4 Standard grade distribution in educational settings as a distribution function over a Quanto-Geometric coordinate system. The s variable grades student's cognitive distal ability and the q variable student count percentile.

Let us now examine the left side of the distribution, namely the distribution of grades F, D, and C-. These students experienced the negative or least performing side of the ranking, which corresponds to the negative side of the position space spread axis. While on the positive side of the axis the position space spread showed the quality of cognitive *absorption*, on the negative side the distal variable equates cognitive *distraction* rather, two qualities that are concordantly antithetic to one another.

A.- The F students have equal amount of separation with the subject as the A students, but in this case separation has the negative quality of *distraction* or attention "away" from the subject, as opposed to attention "toward" the subject experienced by the A students.

B.- At the other end of the spectrum, the C- students could have spent equally long hours of study as the C+ student counterparts demonstrating to be equally studious, but their attention was directed away from the subject for the most part, running in a mode of distraction instead of absorption. Therefore, they score the lesser part of the average grade, a C⁻ grade.

C.- The intermediate group of students in this rubrique are those obtaining the D grade, counterparts of the B grade students. Their D grade is due to a cognitive *distal distractive* quality that sits in between the C⁻ group and the F group. They are essentially marked by the Quadratic ontology experienced in the spectrum at level 4, where position space spread equals scalar density in covariant influence.

The same remarks that we have advanced regarding the granularity of grade distribution on the positive side of the distal variable with respect to ontological qualities are valid here as well. In all, one ought to consider each individual student with his/her learning materials and the educational events occurring as interactions between the two as an event stack forming a *quanto-geometric* object, to be even more exact a *distributive* object. Consequently the population of students shows the full spectrum of possible quanto-geometric correlations or covariance accessible to all material ontologies as described by the Grand Eigenfunction.

This example virtuously demonstrates **how to use the Grand Eigenfunction in the analysis of macroscopic statistical phenomena, in particular how to model the variant variable, whatever its nature, with the Quanto-Geometric space spread independent variable.** One must endeavor to identify in the sample the element that embodies the spatial spread inherent to the Quanto-Geometric visualization, which is not always quite obvious. Once that physicality is uncovered, we are a long way into unraveling the subjacent quanto-geometric qualities to the distribution that make it materially meaningful or causational beyond the mere probability percentiles.

We shall conclude that the above analysis has made it clear that cleverness and/or success in learning is not in direct proportion with compulsive learning behavior, but instead directly proportional to the learner's cognitive *distal* ability. It is thru the mental ability of detachment and elevation that one is able to scope a subject and establish its possible connections with other familiar physical elements, which then confers grasp and significant absorption of the subject matter. The web of symmetries that one is able to establish from scoping a subject constitutes the most important element of successful learning while setting the basis for deep memorization of the subject elements. At best realization, the learner becomes *transcendentally* one with the subject.

7.10. Classical Statistical Physics v. Statistical Quantum Mechanics

Classical Kinetic Theory, first formulated for gases or fluids and subsequently for solids, studies the behavior of large populations as single whole as well as how the properties of the whole relate to those of the constituent units of the whole. It is interesting to note that the formulation of the theory no longer privileges the probability distribution function directly but its functional derivative.

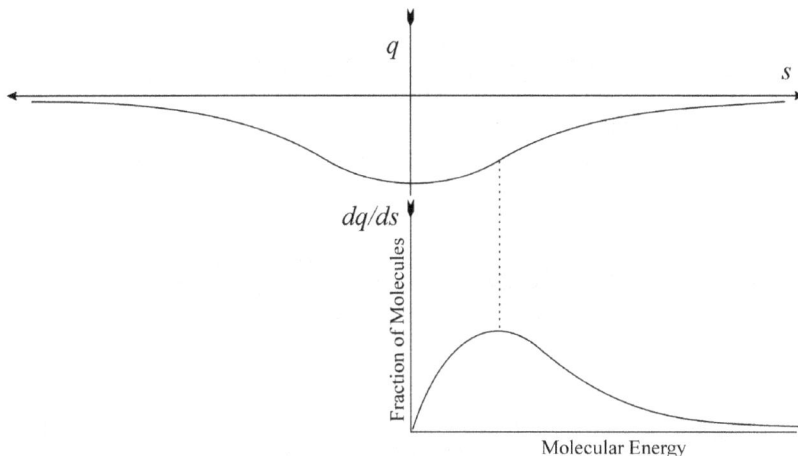

7.5 Distribution of molecular energies in an ideal gas

An important fact to keep in mind is that in practice, I argue theoretically as well, it is almost impossible to study the evolution of the unit particles constituting the whole in a deterministic fashion, essentially along a timeline, that is. The proponents of the theory have argued that the behavior of the unary constituents do not bring pertinent information about the whole population or system, despite the fact that it is the sum total of their properties and behavior that give birth to the macroscopic behavior of the population as a whole. Given the eminent role played by the time variable in classical physics, the proponents of the Kinetic Theory (Maxwell, Boltzmann, etc.) have brushed aside the fact that the behavior of the unit constituents could not be studied deterministically.

We however find a quasi similar scenario at the level of individual particles evolving in a stationary manner conducive to an atomic orbital. In this case the large number count of the population is surrogated by the large number of probabilistic iterations within the position space spread. Furthermore, in quantum mechanics we must assume that the particle is forced to behave as though restrained in a cubic box for the wavefunction to be extolled, which turns out to be a stationary wavefunction independent of time. Likewise, to study the dynamics of a population of gas molecules, we must admittedly assume a cubic container of any desirable volume. The container or its size determines the amplitude of the position space spread but not the space spread itself. Here the role of the position space spread is played by the velocities accessible to the molecules. **It is important to understand that it is the space spread that instantiates the motion which becomes apparent as molecular velocities. It is not the discrete molecule that creates the motion, but its space spread of inhabitation**. The discrete molecules only become submissive to innate spatial motion to be visualized as defoliation or the *un-folding* or *de-foliation* of physical coordinate space. The group of molecules accessible to the many different velocities (or wavelengths of the position-space-spread wavefunction) creates the co-relational discrete scalar states of the distribution. Those discrete groups of molecules are directly comparable to the scalar density states of an electron in orbitalization. To be exact, it is not the number of molecules but the percentile fractional number of molecules that compare to the scalar density states. Effectively the vertical axis of the distribution graph represents a derivative function, by nature a rate. Therefore every velocity is a wavelength value of the material wavefunction of the ideal gas as a whole. The principal figure that characterizes an ideal gas in the Kinetic Theory is the product of the pressure P and volume V occupied by the gas, such that: $PV = \left(\dfrac{2}{3}N\right) \cdot \dfrac{mv^2}{2}$

where v^2 is the square of the average of all molecular velocities, N the total number of molecules and m their individual mass. Note how close v^2 is notionally to probabilistic variance. If we set the moment Q of one molecule to be:

$$Q = mv \text{ , then } PV = \left(\frac{2}{3}N\right) \cdot \left[\frac{Q^2}{2m}\right].$$

PV is thus an expression of the kinetic energy of the gas. In the Hamiltonian operator for the orbitalized or bound electron the core term is:

$$\left[-\frac{\hbar^2}{2m}\right] \cdot \left[\frac{d^2}{dx^2}\right].$$

The first part of the term expresses the kinetic energy displayed by the bound particle. Note its analogy with the same term in the ontological formula for the ideal gas. The difference between $\hbar$ and Q is simply the difference between quasi relativistic motion and non-relativistic classical motion. There is a factor of physical scale as well to account for the difference between the both phenomena visible in the mathematical basis for their formal treatment. One is in development in the atomic Shell of matter and the other in the Astral Shell, medium of inhabitation of the living where we setup our laboratories. Analysis conducted in Chapter 5 has shown the determinants to the formation of those Shells. However our principal interest here is to make explicit the large similarities between the two phenomena in their statistical interpretation.

Once we understand that mass behaves as a density scalar state, that the fraction of gas molecules constitutes a discrete derivative scalar state, coupled with an understanding of the innate dynamical behavior of real position space spread, we will begin to understand a common ontological ultra-structure to all probabilistic behavior in all realms of matter, whether quantum or large-scale. What lies beyond is an equal understanding of the cova-riant relationship between those two ubiquitous variables in the serial development of Shells of matter, well developed throughout previous Chapters of this study.

7.11. Randomness is Relative

Randomness is incomplete in scope. Beyond randomness is the certainty or teleology of types otherwise expressed as normative typologies involving co-variance between the fundamental variables underpinning the ontology of physical objects or evolution of events. Beyond randomness lies the new physics or the physics of:

- Spatial wavefunction of a single object
- Symmetry roots belying all transformation groups
- Ontological covariant spectrum

The incompleteness of Quantum Mechanics does not altogether reside in Quantum Me-chanics itself but essentially in the mathematical statistical theory that sustains its devel-opment. The required analytic categories for a physical interpretation of the ontological configuration of a bound electron simply falter in the probabilistic science of chances as known thus far.

A. Einstein famously conceded that *he cannot imagine the electron hoping like a bug* as implied in the quantum treatment. Could he have conceived that free will is not random either? There is nothing in existence, I repeat NOTHING, that can escape the Quanto-Geometric ontological covariant spectrum. Cognition of the living, bug or human, devel-ops in covariant quantum-space phases according to the normal *quanto-geometric* distri-bution harbored by the Grand Eigenfunction, just the same way or in the same *modalities* that an electron evolves within an atomic orbital.

7.12. Conclusion

In the quest for a Unification Theory that single-handedly furnishes a formal and asser-tive description of the entire sweep of the successive Shells of matter, banking on Proba-bility Theory is good intellectual investment. The universal applicability of the Theory, from quantum density states to rolling dice to the distribution of parameters characteriz-ing the living, is an inestimable asset in that endeavor. The timelessness view implicated in the Theory applies across the full spectrum of Shells of matter, calling for a dramatic ban of the familiar but unphysical time variable in macroscopic ontologies and phenome-nology as well. When re-dimensioned under the umbrella of the Quanto-Geometric Grand Eigenfunction, the Gaussian distribution function at the core of Probability Theory acquires discretisation properties and wavefunction descriptive capabilities that place the applicability of quantum science directly in the midst of macroscopic phenomena. Is this not a new day in Quantum Theory!

7.13. Background...

If you should wonder about the author's background and standing for this publication, I shall disclose that I am not a chaired academician but essentially a studious autodidact of long in the physical sciences. I consider myself a theorist, specifically a system theorist, and a philosopher. My scholarship is that of a maverick and a free thinker who pointedly recognizes the importance of physics in every man's understanding of both individual self and collective self, as well as our common habitat otherwise called matter. My foremost ambition is to unravel the cosmic dimensions of every man's life and contribute to set the stage for a cosmic era to the global culture, which by all accounts is the next stage in the evolution of mankind.

POSTFACE

History of Quanto-Geometry

The Quanto-Geometric Theory has been in the published literature since 1995 when the author released his French publication **La Quanto-Géométrie, Tome I, (ISBN 0-9647466-1-1)**. Written perhaps in an overly sophisticated language and in an extremely dense style for certain, frankly difficult at times to the intended audience, that first cast of the theory presented it on limited mathematical ground and more so on philosophical grounds. The author justly considers nowadays that first publication to simply represent the protogenesis of the Quanto-Geometric Theory.

A few years later in May of 1999, the author released **The Human Mind Inside (ISBN 0-9647466-0-2)**, a softcover English title addressing human mind's complex processes in the light of the Quanto-Geometric Model. This text, largely revised in August 2014, is now available in eBook format **(ISBN 9781311207111)** and print format as well **(ISBN 978-0-9647466-3-3)**. A summary of what to become Chapter 2 of the current publication was already presented in the original 1999 text, amplifying the mathematical foundation of the Quanto-Geometric Theory. It is noteworthy that in said publication the notion of irrevocable dismissal of the time variable in the physical description of nature, one of the key signatures of the Theory and a notion that seems to be gaining traction of late in some corners of theoretical physics, was already in full display.

The current publication comes at a moment when a yearning for renewal of perspectives in theoretical physics, amidst a protracted period of rarefied tangible outcome, seems to be all but pervasive. If anything this book brings to the fore the heightened light of that anticipated candle of conceptual renaissance and the physics of problem resolution for certain. At the very least this exposition will help spark creative and fresh debates whilst setting a solid launch pad for the new Physics. The author may be reached at **quantogeometry@gmail.com (www.quantogeometry.org)**.

APPENDIX

I.- Dirac Lagrangian
For the Hydrogen Atom

Dirac's Lagrangian energy density function for the Hydrogen atom:

$$\psi_{nlm}(r,\vartheta,\varphi) = \sqrt{\left(\frac{2}{na_0}\right)^3 \frac{(n-l-1)!}{2n[(n+l)!]}} \bullet e^{-\rho/2} \rho^l L_{n-l-1}^{2l+1}(\rho) \cdot Y_{lm}(\vartheta,\varphi)$$

where r, θ, φ are the radial and angular variables, ρ the sum of the square of 4 sub-functions of these variables or spinor, n the principal quantum number, l the angular momentum quantum number and m the magnetic quantum number and ψ the main wave density function itself.

II.- Table of Physical Constants
NIST CODATA 2010

QUANTITY	VALUE	UNCERTAINTY	UNITS
(220) lattice spacing of silicon	192.015 5714 e-12	0.000 0032 e-12	m
alpha particle mass	6.644 656 75 e-27	0.000 000 29 e-27	kg
alpha particle mass energy equivalent	5.971 919 67 e-10	0.000 000 26 e-10	J
alpha particle mass energy equivalent in MeV	3727.379 240	0.000 082	MeV
alpha particle mass in u	4.001 506 179 125	0.000 000 000 062	u
alpha particle molar mass	4.001 506 179 125 e-3 0.000 000 000 062 e-3 kg mol^-1		
alpha particle-electron mass ratio	7294.299 5361	0.000 0029	
alpha particle-proton mass ratio	3.972 599 689 33	0.000 000 000 36	
Angstrom star	1.000 014 95 e-10	0.000 000 90 e-10	m
atomic mass constant	1.660 538 921 e-27	0.000 000 073 e-27	kg
atomic mass constant energy equivalent	1.492 417 954 e-10	0.000 000 066 e-10	J
atomic mass constant energy equivalent in MeV	931.494 061	0.000 021	MeV
atomic mass unit-electron volt relationship	931.494 061 e6	0.000 021 e6	eV
atomic mass unit-hartree relationship	3.423 177 6845 e7	0.000 000 0024 e7	E_h
atomic mass unit-hertz relationship	2.252 342 7168 e23	0.000 000 0016 e23	Hz
atomic mass unit-inverse meter relationship	7.513 006 6042 e14	0.000 000 0053 e14	m^-1
atomic mass unit-joule relationship	1.492 417 954 e-10	0.000 000 066 e-10	J
atomic mass unit-kelvin relationship	1.080 954 08 e13	0.000 000 98 e13	K
atomic mass unit-kilogram relationship	1.660 538 921 e-27	0.000 000 073 e-27	kg
atomic unit of 1st hyperpolarizability	3.206 361 449 e-53	0.000 000 071 e-53	C^3 m^3 J^-2
atomic unit of 2nd hyperpolarizability	6.235 380 54 e-65	0.000 000 28 e-65	C^4 m^4 J^-3
atomic unit of action	1.054 571 726 e-34	0.000 000 047 e-34	J s
atomic unit of charge	1.602 176 565 e-19	0.000 000 035 e-19	C
atomic unit of charge density	1.081 202 338 e12	0.000 000 024 e12	C m^-3
atomic unit of current	6.623 617 95 e-3	0.000 000 15 e-3	A
atomic unit of electric dipole mom.	8.478 353 26 e-30	0.000 000 19 e-30	C m
atomic unit of electric field	5.142 206 52 e11	0.000 000 11 e11	V m^-1
atomic unit of electric field gradient	9.717 362 00 e21	0.000 000 21 e21	V m^-2
atomic unit of electric polarizability	1.648 777 2754 e-41	0.000 000 0016 e-41	C^2 m^2 J^-1
atomic unit of electric potential	27.211 385 05	0.000 000 60	V
atomic unit of electric quadrupole mom.	4.486 551 331 e-40	0.000 000 099 e-40	C m^2
atomic unit of energy	4.359 744 34 e-18	0.000 000 19 e-18	J
atomic unit of force	8.238 722 78 e-8	0.000 000 36 e-8	N
atomic unit of length	000 000 17 e-10 m		
atomic unit of mag. dipole mom.	1.854 801 936 e-23	0.000 000 041 e-23	J T^-1
atomic unit of mag. flux density	2.350 517 464 e5	0.000 000 052 e5	T
atomic unit of magnetizability	7.891 036 607 e-29	0.000 000 013 e-29	J T^-2

atomic unit of mass	9.109 382 91 e-31	0.000 000 40 e-31	kg
atomic unit of mom.um	1.992 851 740 e-24	0.000 000 088 e-24	kg m s^-1
atomic unit of permittivity	1.112 650 056... e-10 (exact)	F m^-1	
atomic unit of time	000 000 012 e-17 s		
atomic unit of velocity	2.187 691 263 79 e6	0.000 000 000 71 e6	m s^-1
Avogadro constant	6.022 141 29 e23	0.000 000 27 e23	mol^-1
Bohr magneton	927.400 968 e-26	0.000 020 e-26	J T^-1
Bohr magneton in eV/T	5.788 381 8066 e-5	0.000 000 0038 e-5	eV T^-1
Bohr magneton in Hz/T	13.996 245 55 e9	0.000 000 31 e9	Hz T^-1
Bohr magneton in inverse meters per tesla	46.686 4498	0.000 0010	m^-1 T^-1
Bohr magneton in K/T	0.671 713 88	0.000 000 61	K T^-1
Bohr radius	0.529 177 210 92 e-10 0.000 000 017 e-10 m		
Boltzmann constant	1.380 6488 e-23	0.000 0013 e-23	J K^-1
Boltzmann constant in eV/K	8.617 3324 e-5	0.000 0078 e-5	eV K^-1
Boltzmann constant in Hz/K	2.083 6618 e10	0.000 0019 e10	Hz K^-1
Boltzmann constant in inverse meters per kelvin	69.503 476	0.000 063	m^-1 K^-1
characteristic impedance of vacuum	376.730 313 461 ...	(exact)	ohm
classical electron radius	2.817 940 3267 e-15	0.000 000 0027 e-15	m
Compton wavelength	2.426 310 2389 e-12	0.000 000 0016 e-12	m
Compton wavelength over 2 pi	386.159 268 00 e-15	0.000 000 25 e-15	m
conductance quantum	7.748 091 7346 e-5	0.000 000 0025 e-5	S
conventional value of Josephson constant	483 597.9 e9	(exact)	Hz V^-1
conventional value of von Klitzing constant	25 812.807	(exact)	ohm
Cu x unit	1.002 076 97 e-13	0.000 000 28 e-13	m
deuteron g factor	0.857 438 2308	0.000 000 0072	
deuteron mag. mom.	0.433 073 489 e-26	0.000 000 010 e-26	J T^-1
deuteron mag. mom. to Bohr magneton ratio	0.466 975 4556 e-3	0.000 000 0039 e-3	
deuteron mag. mom. to nuclear magneton ratio	0.857 438 2308	0.000 000 0072	
deuteron mass	3.343 583 48 e-27	0.000 000 15 e-27	kg
deuteron mass energy equivalent	3.005 062 97 e-10	0.000 000 13 e-10	J
deuteron mass energy equivalent in MeV	1875.612 859	0.000 041	MeV
deuteron mass in u	2.013 553 212 712	0.000 000 000 077	u
deuteron molar mass	2.013 553 212 712 e-3 0.000 000 000 077 e-3 kg mol^-1		
deuteron rms charge radius	2.1424 e-15	0.0021 e-15	m
deuteron-electron mag. mom. ratio	-4.664 345 537 e-4	0.000 000 039 e-4	
deuteron-electron mass ratio	3670.482 9652	0.000 0015	
deuteron-neutron mag. mom. ratio	-0.448 206 52	0.000 000 11	
deuteron-proton mag. mom. ratio	0.307 012 2070	0.000 000 0024	
deuteron-proton mass ratio	1.999 007 500 97	0.000 000 000 18	
electric constant	8.854 187 817... e-12 (exact)	F m^-1	
electron charge to mass quotient	-1.758 820 088 e11	0.000 000 039 e11	C kg^-1
electron g factor	-2.002 319 304 361 53 0.000 000 000 53		
electron gyromag. ratio	1.760 859 708 e11	0.000 000 039 e11	s^-1 T^-1
electron gyromag. ratio over 2 pi	28 024.952 66	0.000 62	MHz T^-1
electron mag. mom.	-928.476 430 e-26	0.000 021 e-26	J T^-1
electron mag. mom. anomaly	1.159 652 180 76 e-3	0.000 000 000 27 e-3	
electron mag. mom. to Bohr magneton ratio	-1.001 159 652 180 76 0.000 000 000 27		
electron mag. mom. to nuclear magneton ratio	-1838.281 970 90	0.000 000 75	
electron mass	9.109 382 91 e-31	0.000 000 40 e-31	kg
electron mass energy equivalent	8.187 105 06 e-14	0.000 000 36 e-14	J
electron mass energy equivalent in MeV	0.510 998 928	0.000 000 011	MeV
electron mass in u	5.485 799 0946 e-4	0.000 000 0022 e-4	u

electron molar mass	5.485 799 0946 e-7	0.000 000 0022 e-7	kg mol^-1
electron to alpha particle mass ratio	1.370 933 555 78 e-4	0.000 000 000 55 e-4	
electron to shielded helion mag. mom. ratio	864.058 257	0.000 010	
electron to shielded proton mag. mom. ratio	-658.227 5971	0.000 0072	
electron volt	1.602 176 565 e-19	0.000 000 035 e-19	J
electron volt-atomic mass unit relationship	1.073 544 150 e-9	0.000 000 024 e-9	u
electron volt-hartree relationship	3.674 932 379 e-2	0.000 000 081 e-2	E_h
electron volt-hertz relationship	2.417 989 348 e14	0.000 000 053 e14	Hz
electron volt-inverse meter relationship	8.065 544 29 e5	0.000 000 18 e5	m^-1
electron volt-joule relationship	1.602 176 565 e-19	0.000 000 035 e-19	J
electron volt-kelvin relationship	1.160 4519 e4	0.000 0011 e4	K
electron volt-kilogram relationship	1.782 661 845 e-36	0.000 000 039 e-36	kg
electron-deuteron mag. mom. ratio	-2143.923 498	0.000 018	
electron-deuteron mass ratio	2.724 437 1095 e-4	0.000 000 0011 e-4	
electron-helion mass ratio	1.819 543 0761 e-4	0.000 000 0017 e-4	
electron-muon mag. mom. ratio	206.766 9896	0.000 0052	
electron-muon mass ratio	4.836 331 66 e-3	0.000 000 12 e-3	
electron-neutron mag. mom. ratio	960.920 50	0.000 23	
electron-neutron mass ratio	5.438 673 4461 e-4	0.000 000 0032 e-4	
electron-proton mag. mom. ratio	-658.210 6848	0.000 0054	
electron-proton mass ratio	5.446 170 2178 e-4	0.000 000 0022 e-4	
electron-tau mass ratio	2.875 92 e-4	0.000 26 e-4	
electron-triton mass ratio	1.819 200 0653 e-4	0.000 000 0017 e-4	
elementary charge	1.602 176 565 e-19	0.000 000 035 e-19	C
elementary charge over h	2.417 989 348 e14	0.000 000 053 e14	A J^-1
Faraday constant	96 485.3365	0.0021	C mol^-1
Faraday constant for conventional electric current	96 485.3321	0.0043	C_90 mol^-1
Fermi coupling constant	1.166 364 e-5	0.000 005 e-5	GeV^-2
fine-structure constant	7.297 352 5698 e-3	0.000 000 0024 e-3	
first radiation constant	3.741 771 53 e-16	0.000 000 17 e-16	W m^2
first radiation constant for spectral radiance	1.191 042 869 e-16	0.000 000 053 e-16	W m^2 sr^-1
Hartree energy	4.359 744 34 e-18	0.000 000 19 e-18	J
Hartree energy in eV	27.211 385 05	0.000 000 60	eV
hartree-atomic mass unit relationship	2.921 262 3246 e-8	0.000 000 0021 e-8	u
hartree-electron volt relationship	27.211 385 05	0.000 000 60	eV
hartree-hertz relationship	6.579 683 920 729 e15 0.000 000 000 033 e15		Hz
hartree-inverse meter relationship	2.194 746 313 708 e7	0.000 000 000 011 e7	m^-1
hartree-joule relationship	4.359 744 34 e-18	0.000 000 19 e-18	J
hartree-kelvin relationship	3.157 7504 e5	0.000 0029 e5	K
hartree-kilogram relationship	4.850 869 79 e-35	0.000 000 21 e-35	kg
helion g factor	-4.255 250 613	0.000 000 050	
helion mag. mom.	-1.074 617 486 e-26	0.000 000 027 e-26	J T^-1
helion mag. mom. to Bohr magneton ratio	-1.158 740 958 e-3	0.000 000 014 e-3	
helion mag. mom. to nuclear magneton ratio	-2.127 625 306	0.000 000 025	
helion mass	5.006 412 34 e-27	0.000 000 22 e-27	kg
helion mass energy equivalent	4.499 539 02 e-10	0.000 000 20 e-10	J
helion mass energy equivalent in MeV	2808.391 482	0.000 062	MeV
helion mass in u	3.014 932 2468	0.000 000 0025	u
helion molar mass	3.014 932 2468 e-3	0.000 000 0025 e-3	kg mol^-1
helion-electron mass ratio	5495.885 2754	0.000 0050	
helion-proton mass ratio	2.993 152 6707	0.000 000 0025	

hertz-atomic mass unit relationship	4.439 821 6689 e-24	0.000 000 0031 e-24	u
hertz-electron volt relationship	4.135 667 516 e-15	0.000 000 091 e-15	eV
hertz-hartree relationship	1.519 829 846 0045 e-16 0.000 000 000 0076 e-16 E_h		
hertz-inverse meter relationship	3.335 640 951... e-9	(exact)	m^-1
hertz-joule relationship	6.626 069 57 e-34	0.000 000 29 e-34	J
hertz-kelvin relationship	4.799 2434 e-11	0.000 0044 e-11	K
hertz-kilogram relationship	7.372 496 68 e-51	0.000 000 33 e-51	kg
inverse fine-structure constant	137.035 999 074	0.000 000 044	
inverse meter-atomic mass unit relationship	1.331 025 051 20 e-15 0.000 000 000 94 e-15 u		
inverse meter-electron volt relationship	1.239 841 930 e-6	0.000 000 027 e-6	eV
inverse meter-hartree relationship	4.556 335 252 755 e-8 0.000 000 000 023 e-8 E_h		
inverse meter-hertz relationship	299 792 458	(exact)	Hz
inverse meter-joule relationship	1.986 445 684 e-25	0.000 000 088 e-25	J
inverse meter-kelvin relationship	1.438 7770 e-2	0.000 0013 e-2	K
inverse meter-kilogram relationship	2.210 218 902 e-42	0.000 000 098 e-42	kg
inverse of conductance quantum	12 906.403 7217	0.000 0042	ohm
Josephson constant	483 597.870 e9	0.011 e9	Hz V^-1
joule-atomic mass unit relationship	6.700 535 85 e9	0.000 000 30 e9	u
joule-electron volt relationship	6.241 509 34 e18	0.000 000 14 e18	eV
joule-hartree relationship	2.293 712 48 e17	0.000 000 10 e17	E_h
joule-hertz relationship	1.509 190 311 e33	0.000 000 067 e33	Hz
joule-inverse meter relationship	5.034 117 01 e24	0.000 000 22 e24	m^-1
joule-kelvin relationship	7.242 9716 e22	0.000 0066 e22	K
joule-kilogram relationship	1.112 650 056... e-17 (exact)	kg	
kelvin-atomic mass unit relationship	9.251 0868 e-14	0.000 0084 e-14	u
kelvin-electron volt relationship	8.617 3324 e-5	0.000 0078 e-5	eV
kelvin-hartree relationship	3.166 8114 e-6	0.000 0029 e-6	E_h
kelvin-hertz relationship	2.083 6618 e10	0.000 0019 e10	Hz
kelvin-inverse meter relationship	69.503 476	0.000 063	m^-1
kelvin-joule relationship	1.380 6488 e-23	0.000 0013 e-23	J
kelvin-kilogram relationship	1.536 1790 e-40	0.000 0014 e-40	kg
kilogram-atomic mass unit relationship	6.022 141 29 e26	0.000 000 27 e26	u
kilogram-electron volt relationship	5.609 588 85 e35	0.000 000 12 e35	eV
kilogram-hartree relationship	2.061 485 968 e34	0.000 000 091 e34	E_h
kilogram-hertz relationship	1.356 392 608 e50	0.000 000 060 e50	Hz
kilogram-inverse meter relationship	4.524 438 73 e41	0.000 000 20 e41	m^-1
kilogram-joule relationship	8.987 551 787... e16	(exact)	J
kilogram-kelvin relationship	6.509 6582 e39	0.000 0059 e39	K
lattice parameter of silicon	543.102 0504 e-12	0.000 0089 e-12	m
Loschmidt constant (273.15 K, 100 kPa)	2.651 6462 e25	0.000 0024 e25	m^-3
Loschmidt constant (273.15 K, 101.325 kPa)	2.686 7805 e25	0.000 0024 e25	m^-3
mag. constant	12.566 370 614... e-7 (exact)	N A^-2	
mag. flux quantum	2.067 833 758 e-15	0.000 000 046 e-15	Wb
Mo x unit	1.002 099 52 e-13	0.000 000 53 e-13	m
molar gas constant	8.314 4621	0.000 0075	J mol^-1 K^-1
molar mass constant	1 e-3	(exact)	kg mol^-1
molar mass of carbon-12	12 e-3	(exact)	kg mol^-1
molar Planck constant	3.990 312 7176 e-10	0.000 000 0028 e-10	J s mol^-1
molar Planck constant times c	0.119 626 565 779	0.000 000 000 084	J m mol^-1
molar volume of ideal gas (273.15 K, 100 kPa)	22.710 953 e-3	0.000 021 e-3	m^3 mol^-1

molar volume of ideal gas (273.15 K, 101.325 kPa)	22.413 968 e-3	0.000 020 e-3	m^3 mol^-1
molar volume of silicon	12.058 833 01 e-6	0.000 000 80 e-6	m^3 mol^-1
muon Compton wavelength	11.734 441 03 e-15	0.000 000 30 e-15	m
muon Compton wavelength over 2 pi	1.867 594 294 e-15	0.000 000 047 e-15	m
muon g factor	-2.002 331 8418	0.000 000 0013	
muon mag. mom.	-4.490 448 07 e-26	0.000 000 15 e-26	J T^-1
muon mag. mom. anomaly	1.165 920 91 e-3	0.000 000 63 e-3	
muon mag. mom. to Bohr magneton ratio	-4.841 970 44 e-3	0.000 000 12 e-3	
muon mag. mom. to nuclear magneton ratio	-8.890 596 97	0.000 000 22	
muon mass	1.883 531 475 e-28	0.000 000 096 e-28	kg
muon mass energy equivalent	1.692 833 667 e-11	0.000 000 086 e-11	J
muon mass energy equivalent in MeV	105.658 3715	0.000 0035	MeV
muon mass in u	0.113 428 9267	0.000 000 0029	u
muon molar mass	0.113 428 9267 e-3	0.000 000 0029 e-3	kg mol^-1
muon-electron mass ratio	206.768 2843	0.000 0052	
muon-neutron mass ratio	0.112 454 5177	0.000 000 0028	
muon-proton mag. mom. ratio	-3.183 345 107	0.000 000 084	
muon-proton mass ratio	0.112 609 5272	0.000 000 0028	
muon-tau mass ratio	5.946 49 e-2	0.000 54 e-2	
natural unit of action	1.054 571 726 e-34	0.000 000 047 e-34	J s
natural unit of action in eV s	6.582 119 28 e-16	0.000 000 15 e-16	eV s
natural unit of energy	8.187 105 06 e-14	0.000 000 36 e-14	J
natural unit of energy in MeV	0.510 998 928	0.000 000 011	MeV
natural unit of length	386.159 268 00 e-15	0.000 000 25 e-15	m
natural unit of mass	9.109 382 91 e-31	0.000 000 40 e-31	kg
natural unit of mom.um	2.730 924 29 e-22	0.000 000 12 e-22	kg m s^-1
natural unit of mom.um in MeV/c	0.510 998 928	0.000 000 011	MeV/c
natural unit of time	1.288 088 668 33 e-21 0.000 000 000 83 e-21 s		
natural unit of velocity	299 792 458	(exact)	m s^-1
neutron Compton wavelength	1.319 590 9068 e-15	0.000 000 0011 e-15	m
neutron Compton wavelength over 2 pi	0.210 019 415 68 e-15 0.000 000 000 17 e-15 m		
neutron g factor	-3.826 085 45	0.000 000 90	
neutron gyromag. ratio	1.832 471 79 e8	0.000 000 43 e8	s^-1 T^-1
neutron gyromag. ratio over 2 pi	29.164 6943	0.000 0069	MHz T^-1
neutron mag. mom.	-0.966 236 47 e-26	0.000 000 23 e-26	J T^-1
neutron mag. mom. to Bohr magneton ratio	-1.041 875 63 e-3	0.000 000 25 e-3	
neutron mag. mom. to nuclear magneton ratio	-1.913 042 72	0.000 000 45	
neutron mass	1.674 927 351 e-27	0.000 000 074 e-27	kg
neutron mass energy equivalent	1.505 349 631 e-10	0.000 000 066 e-10	J
neutron mass energy equivalent in MeV	939.565 379	0.000 021	MeV
neutron mass in u	1.008 664 916 00	0.000 000 000 43	u
neutron molar mass	1.008 664 916 00 e-3	0.000 000 000 43 e-3	kg mol^-1
neutron to shielded proton mag. mom. ratio	-0.684 996 94	0.000 000 16	
neutron-electron mag. mom. ratio	1.040 668 82 e-3	0.000 000 25 e-3	
neutron-electron mass ratio	1838.683 6605	0.000 0011	
neutron-muon mass ratio	8.892 484 00	0.000 000 22	
neutron-proton mag. mom. ratio	-0.684 979 34	0.000 000 16	
neutron-proton mass difference	2.305 573 92 e-30	0.000 000 76 e-30	
neutron-proton mass difference energy equivalent	2.072 146 50 e-13	0.000 000 68 e-13	
neutron-proton mass difference energy equivalent in MeV	1.293 332 17	0.000 000 42	

neutron-proton mass difference in u	0.001 388 449 19	0.000 000 000 45	
neutron-proton mass ratio	1.001 378 419 17	0.000 000 000 45	
neutron-tau mass ratio	0.528 790	0.000 048	
Newtonian constant of gravitation	6.673 84 e-11	0.000 80 e-11	m^3 kg^-1 s^-2
Newtonian constant of gravitation over h-bar c	6.708 37 e-39	0.000 80 e-39	(GeV/c^2)^-2
nuclear magneton	5.050 783 53 e-27	0.000 000 11 e-27	J T^-1
nuclear magneton in eV/T	3.152 451 2605 e-8	0.000 000 0022 e-8	eV T^-1
nuclear magneton in inverse meters per tesla	2.542 623 527 e-2	0.000 000 056 e-2	m^-1 T^-1
nuclear magneton in K/T	3.658 2682 e-4	0.000 0033 e-4	K T^-1
nuclear magneton in MHz/T	7.622 593 57	0.000 000 17	MHz T^-1
Planck constant	6.626 069 57 e-34	0.000 000 29 e-34	J s
Planck constant in eV s	4.135 667 516 e-15	0.000 000 091 e-15	eV s
Planck constant over 2 pi	1.054 571 726 e-34	0.000 000 047 e-34	J s
Planck constant over 2 pi in eV s	6.582 119 28 e-16	0.000 000 15 e-16	eV s
Planck constant over 2 pi times c in MeV fm	197.326 9718	0.000 0044	MeV fm
Planck length	1.616 199 e-35	0.000 097 e-35	m
Planck mass	2.176 51 e-8	0.000 13 e-8	kg
Planck mass energy equivalent in GeV	1.220 932 e19	0.000 073 e19	GeV
Planck temperature	1.416 833 e32	0.000 085 e32	K
Planck time	5.391 06 e-44	0.000 32 e-44	s
proton charge to mass quotient	9.578 833 58 e7	0.000 000 21 e7	C kg^-1
proton Compton wavelength	1.321 409 856 23 e-15 0.000 000 000 94 e-15 m		
proton Compton wavelength over 2 pi	0.210 308 910 47 e-15 0.000 000 000 15 e-15 m		
proton g factor	5.585 694 713	0.000 000 046	
proton gyromag. ratio	2.675 222 005 e8	0.000 000 063 e8	s^-1 T^-1
proton gyromag. ratio over 2 pi	42.577 4806	0.000 0010	MHz T^-1
proton mag. mom.	1.410 606 743 e-26	0.000 000 033 e-26	J T^-1
proton mag. mom. to Bohr magneton ratio	1.521 032 210 e-3	0.000 000 012 e-3	
proton mag. mom. to nuclear magneton ratio	2.792 847 356	0.000 000 023	
proton mag. shielding correction	25.694 e-6	0.014 e-6	
proton mass	1.672 621 777 e-27	0.000 000 074 e-27	kg
proton mass energy equivalent	1.503 277 484 e-10	0.000 000 066 e-10	J
proton mass energy equivalent in MeV	938.272 046	0.000 021	MeV
proton mass in u	1.007 276 466 812	0.000 000 000 090	u
proton molar mass	1.007 276 466 812 e-3 0.000 000 000 090 e-3 kg mol^-1		
proton rms charge radius	0.8775 e-15	0.0051 e-15	m
proton-electron mass ratio	1836.152 672 45	0.000 000 75	
proton-muon mass ratio	8.880 243 31	0.000 000 22	
proton-neutron mag. mom. ratio	-1.459 898 06	0.000 000 34	
proton-neutron mass ratio	0.998 623 478 26	0.000 000 000 45	
proton-tau mass ratio	0.528 063	0.000 048	
quantum of circulation	3.636 947 5520 e-4	0.000 000 0024 e-4	m^2 s^-1
quantum of circulation times 2	7.273 895 1040 e-4	0.000 000 0047 e-4	m^2 s^-1
Rydberg constant	10 973 731.568 539	0.000 055	m^-1
Rydberg constant times c in Hz	3.289 841 960 364 e15 0.000 000 017 e15 Hz		
Rydberg constant times hc in eV	13.605 692 53	0.000 000 30	eV
Rydberg constant times hc in J	2.179 872 171 e-18	0.000 000 096 e-18	J
Sackur-Tetrode constant (1 K, 100 kPa)	-1.151 7078	0.000 0023	
Sackur-Tetrode constant (1 K, 101.325 kPa)	-1.164 8708	0.000 0023	
second radiation constant	1.438 7770 e-2	0.000 0013 e-2	m K

shielded helion gyromag. ratio	2.037 894 659 e8	0.000 000 051 e8	s^-1 T^-1
shielded helion gyromag. ratio over 2 pi	32.434 100 84	0.000 000 81	MHz T^-1
shielded helion mag. mom.	-1.074 553 044 e-26	0.000 000 027 e-26	J T^-1
shielded helion mag. mom. to Bohr magneton ratio	-1.158 671 471 e-3	0.000 000 014 e-3	
shielded helion mag. mom. to nuclear magneton ratio	-2.127 497 718	0.000 000 025	
shielded helion to proton mag. mom. ratio	-0.761 766 558	0.000 000 011	
shielded helion to shielded proton mag. mom. ratio	-0.761 786 1313	0.000 000 0033	
shielded proton gyromag. ratio	2.675 153 268 e8	0.000 000 066 e8	s^-1 T^-1
shielded proton gyromag. ratio over 2 pi	42.576 3866	0.000 0010	MHz T^-1
shielded proton mag. mom.	1.410 570 499 e-26	0.000 000 035 e-26	J T^-1
shielded proton mag. mom. to Bohr magneton ratio	1.520 993 128 e-3	0.000 000 017 e-3	
shielded proton mag. mom. to nuclear magneton ratio	2.792 775 598	0.000 000 030	
speed of light in vacuum	299 792 458	(exact)	m s^-1
standard acceleration of gravity	9.806 65	(exact)	m s^-2
standard atmosphere	101 325	(exact)	Pa
standard-state pressure	100 000	(exact)	Pa
Stefan-Boltzmann constant	5.670 373 e-8	0.000 021 e-8	W m^-2 K^-4
tau Compton wavelength	0.697 787 e-15	0.000 063 e-15	m
tau Compton wavelength over 2 pi	0.111 056 e-15	0.000 010 e-15	m
tau mass	3.167 47 e-27	0.000 29 e-27	kg
tau mass energy equivalent	2.846 78 e-10	0.000 26 e-10	J
tau mass energy equivalent in MeV	1776.82	0.16	MeV
tau mass in u	1.907 49	0.000 17	u
tau molar mass	1.907 49 e-3	0.000 17 e-3	kg mol^-1
tau-electron mass ratio	3477.15	0.31	
tau-muon mass ratio	16.8167	0.0015	
tau-neutron mass ratio	1.891 11	0.000 17	
tau-proton mass ratio	1.893 72	0.000 17	
Thomson cross section	0.665 245 8734 e-28	0.000 000 0013 e-28	m^2
triton g factor	5.957 924 896	0.000 000 076	
triton mag. mom.	1.504 609 447 e-26	0.000 000 038 e-26	J T^-1
triton mag. mom. to Bohr magneton ratio	1.622 393 657 e-3	0.000 000 021 e-3	
triton mag. mom. to nuclear magneton ratio	2.978 962 448	0.000 000 038	
triton mass	5.007 356 30 e-27	0.000 000 22 e-27	kg
triton mass energy equivalent	4.500 387 41 e-10	0.000 000 20 e-10	J
triton mass energy equivalent in MeV	2808.921 005	0.000 062	MeV
triton mass in u	3.015 500 7134	0.000 000 0025	u
triton molar mass	3.015 500 7134 e-3	0.000 000 0025 e-3	kg mol^-1
triton-electron mass ratio	5496.921 5267	0.000 0050	
triton-proton mass ratio	2.993 717 0308	0.000 000 0025	
unified atomic mass unit	1.660 538 921 e-27	0.000 000 073 e-27	kg
von Klitzing constant	25 812.807 4434	0.000 0084	ohm
weak mixing angle	0.2223	0.0021	
Wien frequency displacement law constant	5.878 9254 e10	0.000 0053 e10	Hz K^-1
Wien wavelength displacement law constant	2.897 7721 e-3	0.000 0026 e-3	m K

BIBLIOGRAPHY

Edwards, B., Larson, R., and Hostetler, R. **Calculus of a Single Variable**, D.C. Heath and Company, 1994.

Hawking, S., **Une Brève Histoire du Temps**, Editions J'ai lu, 1989

Jean-Claude, J., **La Quanto-Géométrie, Tome I**, Quantometrix,Inc., 1995.

Richmond, A., **Calculus For Electronics**, McGraw-Hill Book Company, 1972.

Thomas, R., and Rosa, A., **The Analysis and Design of Linear Circuits**, Prentice-Hall, Inc. 1994.

Weber, Robert L et al, **Physique Générale**, Mc Graw Hill, 1967

Wilson, J. and Hawkers, F. F. B., **Laser Principles and Applications**, Prentice Hall, 1987

The Regents of the University of Colorado, **Biological Science: Molecules to Men**, Houghton Mifflin Co, 1968

Halliday, David, Resnick, Robert and Walker, Jearl, **Fundamentals of Physics**, John Wiley and Sons, 1997

Daniels, Farrington, and Albert, Robert A., **Physical Chemistry**, John Wiley and Sons, 1963

Tortora, Gerard J., Funke, Berdell R. and Case, Christine L., **Microbiology An Introduction**, Benjamin Cummings Publishing, 1982

Acosta, Virgilio, Cowan, Clyde L., and Graham, B.J., **Essentials of Modern Physics**, Harper & Row Publishers, 1973

Cruz, Andoni, Chamizo, Jose, Garritz, Diana, **Estructura Atómica: Un Enfoque Químico**, Fondo Educativo Interamericano, 1986

Hobson, Art, **Physics : Concepts and Connections**, Pearson Addison-Wesley, 2010

Rees, Martin, Just Six Numbers: **The Deep Forces That Shape the Universe**, Weidenfeld & Nicolson, 1999

Feynman, Richard, P., **Theory of Fundamental Processes**, W.A. Benjamin, 1962

Feynman, Richard, Pl, **QED: The Strange Theory of Light and Matter**, Princeton University Press, 1985

Born, Max, **Deciphering the Cosmic Number: The Strange Friendship of Wolfgang Pauli and Carl Jung**, W. W. Norton & Co, 2009

Dirac, Paul, M., **The Evolution of the Physicist's Picture of Nature**, Scientific American, May 1963

Peskin, Michael, E., and Schroeder, Daniel, V., **Introduction to Quantum Field Theory, Sarat Book House**, 2005

Markowitz, W., Hall, R., G. and Essen, L., **Frequency of Cesium in Terms of Ephemeris Time**, Physical Review Letters, 1958

Lide, D. R., **A Century of Excellence in Measurements, Standards and Technology**, CRC Press, 2001

Woit, Peter, **Not Even Wrong: The Failure of String Theory and the Search for Unity in Physical Law**, Basic Books, 2006

Charles, Mortimer, E., **Chemistry**, Wadsworth Publishing, 1983

Schrödinger, E., **Structure of Spacetime**, Cambridge University Press, London 1950.

Organisation Inter-gouvernementale de la Convention du Mètre, **The International System of Units (SI),** 8th Edition, 2006

Smolin, Lee, **The Trouble with Physics: The Rise of String Theory, the Fall of a Science, What Comes Next**, Mariner Books, 2007

George B. Arfken & Hans J. Weber, **Mathematical Methods for Physicists**, 6th edition, Academic Press, 2005

Giovanni Sansone (translated by Ainsley H. Diamond), **Orthogonal Functions**, Interscience Publishers, 1959

Courant, R.; Hilbert, D. **Methods of Mathematical Physics**

Katznelson, Y., **An Introduction to Harmonic Analysis**, 3rd Edition. Cambridge University Press, 2004

Sunder, V.S. **Functional Analysis: Spectral Theory**, Birkhäuser Verlag, 1997

P.W. Atkins, **Quanta: A handbook of concepts**, Oxford University Press, 1974

Farwaz T. Ulabi, Michel M. Maharbiz, **Circuits**, National Technology and Science Press, 2009

Peleg, Y., Pnini, R.; Zaarur, E.; Hecht, E., **Quantum Mechanics** (Schaum's Outline Series), 2nd Edition, McGraw Hill, 2010

H. McKean and V. Moll, **Elliptic Curves,** Cambridge University Press, 1999

Hedrick, Philip W., **Population Biology: The Evolution and Ecology of Populations**, Jones and Bartlett Publishers, 1984

INDEX

www.ingramcontent.com/pod-product-compliance
Lightning Source LLC
Chambersburg PA
CBHW060353220326

41598CB00023B/2901